ELEMENTS

OF THE

DIFFERENTIAL AND INTEGRAL

CALCULUS,

BY

WILLIAM SMYTH, A. M.,

PROFESSOR OF MATHEMATICS IN BOWDOIN COLLEGE.

Second Edition.

PORTLAND,

PUBLISHED BY SANBORN & CARTER.

PRESS OF J. GRIFFIN, BRUNSWICK.

1859.

PREFACE TO THE FIRST EDITION.

THE principles of the Calculus are reached from three different points of view in the methods respectively of Leibnitz, Newton, and Lagrange. The method of the latter not being adapted to elementary instruction, the choice for this purpose lies between those of the two former. The recent text-books, both English and French, are in general based upon the method of Newton. The expediency of this may well be questioned. The artifice which lies at the basis of the Calculus, consists in the employment of certain special auxiliary quantities adapted to facilitate the formation of the equations of a problem. The limit, or differential coefficient, the auxiliary employed in the method of Newton, is not easily represented to the mind, and being composed of two parts which cannot be separately considered, it is with the more difficulty applied to the solution of problems. On the other hand the differential, the auxiliary employed in the method of Leibnitz, is simple in itself, is very readily conceived, and adapts itself with wonderful facility to all the different classes of questions which require for their solution the aid of the Calculus. From these and similar considerations, confirmed by the results of experience, I do not hesitate to prefer the method of Leibnitz as the basis of an elementary work on the Calculus. A perfect knowledge of the subject requires, indeed, that it should be examined from all the different points of view which have been named. But the learner should begin with the method of Leibnitz. In accordance with these views the present work has been prepared. The following analysis exhibits the course pursued.

The first object is to show the necessity for the new instrument of investigation we are about to examine. A simple problem is, therefore, introduced for which an equation cannot be directly obtained in terms of the quantities which enter into it, and which require to be considered. In these circumstances the ordinary expedient of analysis, is the employment of auxiliary quantities to aid in the formation of the equation required. But from the variable nature of the quantities which the problem presents, the auxiliaries most suitable to be employed, it is obvious, are not the finite and determinate quantities of ordinary algebra, but certain indeterminate quantities capable of being taken as small as we please, without changing thereby the values of the quantities for which they are to be substituted or in connection with which they are to be used. We are thus led to the consideration of quantities infinitely small, or infinitesimals; which form the subject of the first section.

The auxiliaries employed in any case must, it is evident, sustain relations to their primitives, in virtue of which they may be eliminated, after they have served the purpose for which they are introduced. We have, then, next to establish such relations for the new auxiliaries we wish to employ. The variable quantities of the problem may be regarded as changing their values by increments or decrements infinitely small. If we take, then, the difference between any two successive states of the variables, the results will be quantities infinitely small, which will have also a necessary relation to the primitives from which they are derived. We may employ them, therefore, for the auxiliaries required. They are called differentials; and the process by which they are obtained is called differentiation; the rules for which, for simple algebraic quantities, are developed in the second section.

We have now our instrument in part, and proceed at once to its use. We commence with the problem of tangents, one of the most interesting of the problems which occupied the attention of the ablest geometers of antiquity, and which long defied their efforts and baffled their skill. We find it yield at once to the instrumentality we have now acquired. The third section is occupied with various particular cases of this general problem, and the student is thus introduced to the Differential Calculus, of the nature of which he is now enabled to form a general idea.

In the problems solved thus far, the quantities whose relations are sought enter directly into the differential equations obtained. We are able, therefore, to eliminate by the ordinary processes of algebra, the auxiliaries employed. But we meet next with a class of problems in which this cannot be done. And in which the only mode of freeing the equations from the indeterminates employed consists in returning from them to their primitives, by a process the reverse of that by which they are obtained.

The infinitely small quantities we have employed as auxiliaries, may be regarded as the elements of which the variable quantities of the problem are respectively composed. This reverse process will then consist in the summing up or taking the whole of these elements. It is called integration, the rules for which, for simple algebraic quantities, are developed in the fourth section. Proceeding immediately to the application of our instrument as now improved, we introduce in the fifth section various problems upon the quadrature of areas, and the cubature of solids; the solution of which gives the student an elementary view of the Integral Calculus.

The problems of the rectification of curve lines, and the quadrature of the surfaces of the solids of revolution, requiring the aid of both the processes of elimination already employed, brings into use the entire instrument. It is thus seen as a whole, and the student has now a general idea of THE CALCULUS, as the entire instrument by way of eminence is called. What remains, is only its more full development, with such new applications as will most clearly exhibit its power.

With this object in view, we pass to the sixth section. A proposed problem being too difficult for the auxiliaries already obtained, we naturally seek for additional ones by a repetition of the artifice already employed. Regarding our first auxiliaries, therefore, as now becoming primitives in their turn, we take their differentials, and again the differentials of these last, and so on. We thus obtain second, third, &c. differentials. These form the subject of the sixth section. Making immediate application of these new auxiliaries, we introduce in the seventh section the problem of the development of a function of one or more variables, and obtain the important theorems of Maclaurin and Taylor. And with resources thus increased, we discuss at large in the eighth and ninth sections the great problem of maxima and minima, and the whole theory of curves.

Thus far our problems have involved algebraic quantities only. Passing next to those which involve transcendental functions, we develop in the tenth section rules both for the differentiation and integration of quantities of this description. The discussion of the sinusoid, the logarithmic curve, and the spirals in general, presents in the eleventh section an application of the Calculus to problems depending upon transcendental quantities.

It must, by this time, have been perceived that the Differential Calculus is far in advance of the Integral. The twelfth section is, therefore, occupied with the development of various artifices and methods of reduction adapted to facilitate the process of integration, or the elimination of the auxiliaries we have occasion to employ. With these increased facilities we advance, in the thirteenth section, to a full examination of the nature and properties of the Cycloid, the most beautiful as well as important of all the transcendental curves. Here the full power of the Calculus is more distinctly exhibited. Problems are solved by it, as with the dash of a pen, which eluded the grasp of the most distinguished of the ancient geometers, or were solved only after laborious efforts, and by methods alike circuitous and complicated.

In the fourteenth section, some additional problems upon the quadrature of areas and cubature of solids are introduced, and with these is closed the application of the Calculus to problems of pure geometry.

In the fifteenth and sixteenth sections we enter a wider field and upon topics of a higher interest. In these the Calculus is applied to various problems of Mechanics involving varied motion, motion along curve lines, and the determination of forces; problems upon equilibrium, the determination of the centre of gravity of bodies, and the pressure and discharge of fluids. We thus see the application of the Calculus to the Physical Sciences, the sphere to which it is specially adapted, and within which its powers have been most remarkably displayed.

At this point, we take our final step in the improvement of the instrument we are considering. We come to a class of problems far transcending in difficulty those which have previously been solved; and which require for their solution a higher degree of indeterminateness in the auxiliaries employed. The new class of auxiliaries required are differentials under a new point of view. They are called variations. Their derivation with some examples of their use forms the subject of the seventeenth section. We close our necessarily limited view of the sublime instrument before us, by its application in the eighteenth section to two of the most important problems of astronomy—the attraction of spheres, and the investigation of the law of the force which binds the planets and comets to their orbits. The work terminates in the nineteenth section, with a brief exposition of the methods of Newton and Lagrange, and a few miscellaneous examples.

The plan of the work differs, it will be perceived, widely from the many excellent text-books on the subject recently published. It is submitted with diffidence in regard to its execution, but with great confidence that the plan itself is well adapted to introduce the student by a natural and easy process to a general knowledge of the Calculus, to enable him clearly to understand in what it consists, and to discern its power; and especially to awaken in him an interest in those profound investigations in respect to which it is the appropriate instrument.

The work completes the course of text-books of pure mathematics prepared by the author. Its materials have been selected from the ordinary sources. The general view taken of the philosophy of the Calculus is the same with that of Carnot in his Reflections on the Metaphysics of the Differential and Integral Calculus, and of Comte in his Positive Philosophy.

WM. SMYTH.

Bowd. Coll., March, 1854.

NOTE TO THE SECOND EDITION.

In the present edition of this work no changes have been made, except, in accordance with its general plan, to adapt it more perfectly to the wants of the Recitation room. The figures have all been inserted with the text, which will be found a great convenience. In the nineteenth section a more full account is given of the method of Newton. Rules for finding the Differential Coefficients of Algebraic quantities are derived by this method, and some examples of their application, in the solution of problems, given. The student will thus have, directly, the means of comparing the method of Newton with that of Leibnitz upon which the present work is based. Students, in our Colleges, generally approach the Calculus with an erroneous impression in respect to the difficulty of the study. This impression, it is believed, is strengthened by the want of due attention to the distinction between the Calculus as a logical instrument, and the subjects to which it is applied. In the former point of view the Calculus presents no special difficulty; and in the latter, very many of its most interesting and important applications require no higher mental effort than that to which the student has already been accustomed, in the study of Algebra to the extent required in a College course. In addition to the use of the method of Leibnitz, the author has kept the distinction in question constantly in view. He has endeavored also to present such a course of topics for the application of the Calculus as fall fully within the range of College instruction, and which, without the aid of this instrument, would be taught by far inferior methods. The several topics discussed are presented so as to advance gradually from the more simple to that which is more difficult. Each topic, moreover, as far as practicable, is presented as a whole, instead of its different parts being scattered over the entire work. Greater unity in the discussions is thus secured, with greater facility for selection, when, from any circumstances, the entire work cannot be used. The present work owes its origin to a want felt by the author in the instruction of his own classes. Its success thus far, has entirely met his expectations. It is now presented as a simple contribution to the means of elementary instruction, in the hope that it may serve to increase the attention now given in our Colleges and higher Seminaries to the important branch of Mathematics of which it treats.

Bowd. Coll., October, 1859.

TABLE OF CONTENTS.

SECTION I.

SECTION II.

SECTION III.

SECTION IV.

SECTION V.

SECTION VI.

SECTION VII.

SECTION VIII.

SECTION IX.

SECTION X.

SECTION XI.

SECTION XII.

SECTION XIII.

SECTION XIV.

SECTION XV.

THE

CALCULUS.

SECTION I.

PRELIMINARY PRINCIPLES.

1. When an equation for a question can be obtained *directly* by means of the quantities which come under consideration, ordinary Algebra is sufficient to furnish the necessary rules for its complete solution. But numerous questions occur, especially in the Physical Sciences, the equations for which cannot thus be obtained, and which, in consequence, require for their solution new expedients or methods of analysis.

2. Let the following simple problem of this description be proposed, viz.

To find the space through which a body, acted upon by gravity, will descend in a given time.

Let s represent the space, t the time, and v the velocity the body has acquired at the end of the time.

Since gravity is an accelerating force, the velocity, it is evident, will vary with the time elapsed, and, conversely, the time will vary with the velocity. This varying character of the time and velocity, both of which must be considered in determining the space, renders it difficult to form *directly* the equation of the problem, and we are thus led to seek some indirect method by which the difficulty may be obviated and the required equation be obtained.

With a little attention it will be seen that if the time be divided into portions each indefinitely small, for any one of these portions the velocity may be regarded as uniform. Let the body, then, be supposed, with the velocity acquried in the time t, to continue its motion during one of these indefinitely small portions of the time, which we will designate by t'. During this time it will describe, it is evident, a corresponding indefinitely small portion of the space, which we will represent by s'. Then, since in a uniform motion the velocity is equal to the space divided by the time, we shall have

$$v = \frac{s'}{t'}. \qquad (1)$$

But gravity, being a constant force, will generate the same velocity in each successive equal interval of time. Thus, if g represent the velocity generated by gravity in the *unit* of time, in 2 units it will be $2g$; and, in general, in t units it will be gt, and we shall have

$$v = gt. \qquad (2)$$

Comparing these equations we obtain

$$s' = gt \times t'. \qquad (3)$$

We have thus obtained an equation for the problem by means of the indefinitely small quantities s' and t' which we have employed as auxiliaries, and which we suppose to be derived from the primitive quantities s and t by some fixed and uniform law. This law being determined, the auxiliaries s', t', it is evident, may be eliminated; and the result will be an equation in s, t, and constants only, from which the value of s may be obtained, and the problem thus be completely solved.

3. We shall pursue the problem no further at present; our object being simply to indicate the circumstances which gave rise to

the new branch of Mathematics we are about to consider, and the peculiar artifice of calculation by which it is characterized.

The problems for the solution of which this branch of the Mathematics was invented, are among the most important that can be presented to the human mind. On this account, as well as from the inherent difficulty of their solution, it has been called, by way of eminence, the CALCULUS.

4. The method of the Calculus is, for our present purpose, sufficiently obvious. It consists in the employment of certain auxiliary quantities adapted, by their relations to each other and the primitive quantities for which they are substituted or in connection with which they are used, to facilitate the formation of the equations necessary to the solution of problems. We have, therefore, at the outset, two questions to consider. 1°. What is the nature of the auxiliary quantities thus employed? 2°. How are these quantities derived from their primitives?

NATURE OF THE AUXILIARY QUANTITIES EMPLOYED IN THE CALCULUS.

5. In order that the quantities s', t', may be employed as auxiliaries in forming an equation for the preceding problem, we must be able, it is evident, while they retain the required relation to each other, to make them as small as we please, without altering thereby the values of the primitive quantities s and t with which they are compared, and the relation of which is sought. It is the capability of being thus indefinitely diminished, which characterizes the quantities employed in the Calculus as auxiliary to the formation of equations. In their nature indeterminate, they are regarded in practice as *infinitely small*, or as *infinitesimals.* And the Calculus, on account of its use of them, is sometimes called the *Infinitesimal Analysis.*

Before proceeding to the second inquiry in respect to these auxiliary quantities, let us examine more particularly what is to be understood by quantities *infinitely great*, or, *infinitely small;* and the subordination which must be regarded as existing among these quantities in calculation.

INFINITY. INFINITESIMALS.

6. We can form no conception of a quantity absolutely infinite. Nor are the terms great and small applicable to quantities in themselves considered, but only in their relation to each other, or to a common standard. One quantity is greater than another when it contains the other more than once. It is *infinitely* great in comparison with another, when *no number can be found sufficiently large to express the ratio between them*, or the number of times one contains the other. Thus x is infinitely great in relation to a, when no number can be found sufficiently large to express the quotient $\frac{x}{a}$. In like manner a is infinitely small in relation to x, when no number can be found sufficiently small to express the quotient $\frac{a}{x}$.

7. But though we cannot assign the value of the ratio $\frac{x}{a}$, this does not prevent our supposing another quantity as large in relation to x, as x is in relation to a; for, whatever the magnitude of x, we may have, it is evident, the proportion

$$a : x :: x : \frac{x^2}{a},$$

in which the fourth term, $\frac{x^2}{a}$, will contain x as many times as x is supposed to contain a.

In like manner we may have the proportion

$$x : a :: a : \frac{a^2}{x}$$

in which, if a is infinitely small in relation to x, $\frac{a^2}{x}$ will be infinitely small in relation to a; for $\frac{a^2}{x}$ is contained as many times in a as a is contained in x.

8. Moreover, we may again suppose quantities infinitely greater, or infinitely smaller, than those last named, and so on without limit.

Quantities of this description are called infinitely great or infinitely small quantities of *different orders.*

Thus if x is infinitely great in comparison with 1, x^2 will be infinitely great in comparison with x, x^3 in comparison with x^2, and so on. The first, or x, is called an infinite quantity of the *first order;* the second, or x^2, an infinite quantity of the *second order*, and so on.

In like manner if $\frac{1}{x}$ is a quantity infinitely small, $\frac{1}{x^2}$ will be infinitely small in comparison with $\frac{1}{x}$, $\frac{1}{x^3}$ in comparison with $\frac{1}{x^2}$, and so on. Thus $\frac{1}{x}$ is an infinitesimal of the *first* order, $\frac{1}{x^2}$ an infinitesimal of the *second* order, and so on.

If we take the series

$$x^3, x^2, x, 1, \frac{1}{x}, \frac{1}{x^2}, \frac{1}{x^3}$$

in which x is infinite, each term, it is evident, will be infinitely large in relation to that which immediately follows, and infinitely small in relation to that which immediately precedes it.

9. It will be observed, 1°, If two quantities x and y are each infinitely small, their product will be an infinitesimal of the *second* order; for we have the proportion

$$1 : x :: y : xy,$$

in which, if x is infinitely small in relation to 1, xy will be infinitely small in relation to y; that is, it will be an infinitesimal of the *second* order. In like manner, z being also an infinitesimal, xyz will be an infinitesimal of the *third* order, and so on.

2°. Though two quantities are each infinitely small, it does not follow that they are equal. Indeed, two infinitely small quantities may have any finite ratio whatever. Thus let $\frac{a}{x}, \frac{b}{x}$ represent two infinitesimals. We shall have, it is evident, the proportion

$$\frac{a}{x} : \frac{b}{x} :: a : b,$$

in which, though $\frac{a}{x}, \frac{b}{x}$ are each indeterminate, that is, may each be made as small as we please, the ratio between them is always determinate, viz. that of $a : b$.

3°. Moreover, if an infinitesimal be multiplied by a finite quantity, its order of infinity will not thereby be changed. Thus $\frac{1}{x}, \frac{a}{x}$ are each infinitesimals of the first order when x is infinite, although the second is a times the first.

4°. If x is infinite in comparison with a, conversely a is an infinitesimal in comparison with x.

10. In the use of infinitesimals, an infinitesimal may be omitted by the side of a finite quantity; and, in general, an infinitesimal of any order whatever by the side of one of an inferior order.

1. Thus, let there be the equation

$$y = a + \frac{b}{x}.$$

Here, if x is infinite, $\frac{b}{x}$ can add nothing to a. Indeed as x increases, the fraction $\frac{b}{x}$ decreases, and may be made as small as we please by increasing x. If then x be greater than any finite or assignable quantity, $\frac{b}{x}$ will be less than any finite or assignable quantity. It can add, therefore, nothing finite to the value of a, or will become as 0 in comparison with it, and may by consequence be rejected.

In like manner, if we have the equation

$$y = a - \frac{b}{x},$$

if x is infinite, $\frac{b}{x}$ can subtract nothing from a. In either case, therefore, if x is infinite, we shall have

$$y = a.$$

2. To find the product of B into $x - \mathrm{A}$, on the hypothesis that x is infinite; or, which is the same thing, that A is an infinitesimal compared with x.

The product of B into $x - \mathrm{A}$ may be put under the form

$$\mathrm{B}x\left(1 - \frac{\mathrm{A}}{x}\right)$$

which, for the reasons above, reduces to $\mathrm{B}x$, when $x = \infty$.

3. To find the value of y in the equation $y = x^2 - ax + b$, when x is infinite.

The right hand member of this equation may be put under the form

$$x^2\left(1 - \frac{a}{x} + \frac{b}{x^2}\right)$$

which reduces to x^2, when $x = \infty$.

4. To find the value of y in the same equation when x is an infinitesimal. Ans. $y = b$.

5. To find the value of the fraction $\frac{3x + a}{5x + b}$; 1°. on the hypothesis $x = \infty$; 2°. on the hypothesis that x is an infinitesimal.

Ans. $\frac{3}{5}$, and $\frac{a}{b}$ respectively.

ADAPTATION OF INFINITESIMALS TO SIMPLIFY THE EQUATIONS OF A PROBLEM.

11. We have thus seen that an infinitely small quantity may always be omitted by the side of a finite quantity, and in general an infinitesimal of any order whatever by the side of one of an inferior order. By reason of the terms which may be omitted, in virtue of this principle, the equations of questions, formed by means of infinitesimals as auxiliaries to the primitive quantities, will be much more simple, and be more easily established. The advantage derived from this simplification will be obvious, when it is recollected how much the difficulty in the solution of a problem is increased, when, for example, it is necessary to introduce into its equations the second, third, and, in general, the higher powers of the quantities to be considered.

12. But the question arises, will not these omissions affect the accuracy of the results at which we arrive in the use of these auxiliaries. In reply to this question we remark,

1°. Admitting that errors occur in consequence of these omissions, they may be made as small as we please, or less than any finite or conceivable magnitude. They must, therefore, ultimately become as 0, or entirely disappear.

2°. These omissions are necessary to impress upon the auxiliaries employed the true character we assign to them.

If a question is solved upon a certain hypothesis in respect to the quantities employed, then, whatever in the course of the operations is contradictory to this hypothesis must be struck out, or the result will not conform to the data. Suppose that, in the solution of a problem, one of the quantities, x for example, is regarded as infinite, and we arrive at the equation

$$y=\frac{3x+a}{5x+b};$$

a and b, it is evident, must be struck out by the side of x. For the supposition that a and b can have any value by the side of x destroys the previous hypothesis, $x = \infty$, upon which the solution of the question depends.

2. As a second illustration let us take the series

$$\frac{1}{2}, \frac{2}{3}, \frac{3}{4}, \frac{4}{5}, \frac{5}{6}, \&c.$$

The terms of this series approach more nearly to unity, as they are farther distant from the beginning. A term, it is evident, can never exceed unity, and will reach this limit only when it is at an infinite distance from the beginning. Let x denote the distance of a term from the beginning; the fraction

$$\frac{x}{x+1}$$

will then represent, generally, any term of the series.

In this expression let $x = \infty$; then, from what has been said, the expression reduces to $\frac{x}{x}$ or 1; which will be the value of the term at an infinite distance from the beginning, as it should be.

The omission, then, of unity by the side of x, when x is supposed to be infinite, only makes the result conform to the hypothesis; and so far from leading to error is what is necessary to make the result rigidly exact.

We thus see that an infinitesimal, when connected by algebraic addition with a finite quantity or with an infinitesimal of an inferior order, not only *may*, but *must* be rejected, in order to secure absolute accuracy.

13. The nature of the quantities adapted to be employed as auxiliaries in the solution of questions, the equations for which it is difficult or impossible to obtain directly, has now been sufficiently explained. To apply these auxiliaries to their purpose, we must be able, as we have seen, to effect their final elimination from the equations in which they have been employed. They must sustain,

therefore, a definite and known relation to their primitives. We come, then, to our second object of inquiry, viz. How are these auxiliary quantities derived from their primitives?

....

SECTION II.

DERIVATION OF THE AUXILIARY QUANTITIES. DIFFERENTIAL CALCULUS.

14. In the questions which occur in the Calculus, as in the Analytic Geometry, there are two classes of quantities which come under consideration,

1°. Quantities which preserve invariably the same value throughout the whole course of the operations in which they are employed. These are called *constants*, and are usually denoted by the first letters of the alphabet, as a, b, c, &c.

2°. Quantities which admit of different values or different degrees of value in the same expression. These are called *variables*, and are usually denoted by the last letters of the alphabet, as x, y, z, &c.

3°. To these must now be added a third class, viz. the infinitesimals employed as auxiliaries. These, although they vary in a certain manner, cannot, as the ordinary variables, have any determinate values assigned to them. They may be characterized as quantities *always variable*, or as *indeterminates*.

15. The quantities which enter into a question, and the relations of which are sought, may all be considered as composed of *elements* infinitely small in comparison with them. These infinitely small elements are the quantities actually employed in the Calculus as auxiliaries to facilitate the establishment of equations. They will necessarily have to one another relations more simple and easy to discover than those of the primitive quantities for which they are substituted or with which they are used. They may, moreover, be

derived from their primitives by very simple rules which we now proceed to develop.

16. A variable quantity may be considered as changing its magnitude, or as having arrived at a certain state of magnitude, by the successive addition or subtraction of the infinitely small quantity regarded as its element. To find, then, the element, or the infinitely small increment or decrement, by which a variable is increased or decreased at each instant, we have only to find the value of the variable at any one instant, and then in the instant next immediately following; or, which is the same thing, we have only to find the values of the variable in any two successive states of its magnitude infinitely near to each other. The difference between these two values will be the element sought. It is called the *differential* of the quantity.

To indicate the element or differential of a simple variable, we write the letter d, signifying *difference of*, before it. Thus dx indicates the differential of x, dy the differential of y. If the variable is raised to a power, we place a point after the d; thus the differential of x^2 is indicated by $d.\,x^2$. In like manner the differential of xy may be indicated by $d.\,xy$.

If the variable is a compound quantity, we inclose it in a parenthesis and place the letter d before it.

Thus the differential of $x+y$ is indicated by $d\,(x+y)$; the differential of x^2-ax by $d\,(x^2-ax)$.

In all cases the letter d is to be regarded in the Calculus as the sign of a differential, and never as a factor of the quantity before which it is placed, or with which it is connected.

The process by which a differential is found is called *differentiation;* and that part of the Calculus which relates to finding the differentials of quantities is called the *Differential Calculus.*

DIFFERENTIATION.

17. Let it be proposed to find the differential of the quantity $x+y-z$ composed of the simple variables x, y, and z.

In order to this, let x, y, and z represent the values of the variables at any one instant; to find their value for the instant immediately following, we suppose x to receive an increment dx, and to become $x + dx$; y in like manner to become $y + dy$; and z to become $z + dz$. We thus have

The second value of the quantity	$= x + dx + y + dy - z - dz.$
From which subtracting the first	$= x \quad + y \quad - z.$
We obtain the differential	$= dx + dy - dz.$

To find, therefore, the differential of a quantity composed of several simple variables of the first degree connected by the signs of addition or subtraction, *we write the letter* d *before each variable, leaving the sign of each unchanged.*

2. To find the differential of the quantity $ax + by$.

We have the second value of the quantity

$$= a(x + dx) + b(y + dy)$$
$$= ax + by + adx + bdy$$

Subtracting the first	$= ax + by$
The differential	$= adx + bdy$

Thus, *if a variable is multiplied or divided by a constant quantity, its differential will be multiplied or divided by the same quantity.*

3. Find the differential of $ax - y + c$.

Second value of the quantity	$= ax - y + adx - dy + c.$
Subtracting the first	$= ax - y + c.$
The differential	$= adx - dy.$

Whence, *if a constant quantity is connected with a variable by the sign of addition or subtraction, the constant will disappear in differentiation.*

This, it is evident, should be the case, since a constant, not being capable of increase or diminution, can have no differential, or must have 0 for its differential.

18. Let it be required next to find the differential of the product xy of two simple variables.

Considering x as becoming $x + dx$, and y as becoming $y + dy$, we have for the second value of the quantity

$$(x + dx)(y + dy) = xy + xdy + ydx + dydx.$$

Subtracting the first $= xy$

Difference $= xdy + ydx + dydx.$

But, since dx and dy are by hypothesis infinitesimals, the product $dxdy$ is an infinitesimal of the second order, and must be omitted by the side of $xdy + ydx$. The differential will be, therefore, $xdy + ydx$.

In like manner, if there are three factors x, y, z, we shall have

$$d(xyz) = xydz + xzdy + yzdx,$$

and so on, for any number of factors.

In general, therefore, to find the differential of a product of several factors, *we find the differential of each variable factor successively, and multiply it by all the rest as if they were a constant co-efficient. The sum of these partial products will be the differential sought.*

19. Let it be proposed next to find the differential of the power of a simple variable, x^2, for example.

For the second value of the quantity we have

$$(x + dx)(x + dx) = x^2 + 2xdx + dx^2.$$

Subtracting the first $= x^2$

Difference $= 2xdx + dx^2.$

But dx^2 is an infinitesimal of the second order and must be omitted by the side of $2xdx$. Hence,

$$d. x^2 = 2xdx.$$

In like manner $d. x^3 = 3x^2dx$.

To take the general case, let it be required to find the differential of x^m. We shall have for the second value of the variable,

$$x^m + mx^{m-1}dx + \frac{m(m-1)}{2}x^{m-2}dx^2 + \&c.$$

From this subtracting the first, and rejecting the terms multiplied by infinitesimals of the second and higher orders, we obtain

$$d.x^m = mx^{m-1}dx.$$

We have, therefore, the following rule by which to find the differential of any power of a variable quantity. *Multiply by the exponent, diminish the exponent by unity, and multiply the result by the differential of the variable.*

20. Let it now be required to find the differential of a fraction $\frac{x}{y}$.

Placing the denominator in the numerator the fraction becomes xy^{-1}. Applying the preceding rule we obtain

$$d.xy^{-1} = y^{-1}dx - xy^{-2}dy,$$

or, restoring the denominators and reducing to a common denominator,

$$d.\frac{x}{y} = \frac{ydx - xdy}{y^2};$$

and we have the following rule by which to find the differential of a fraction. *Multiply the differential of the numerator by the denominator; from the product subtract the differential of the denominator multiplied by the numerator, and divide the remainder by the square of the denominator.*

21. To find the differential of $x^{\frac{1}{2}}$. The rule, art. 19, applies equally to quantities affected with fractional exponents. Thus,

$$d.x^{\frac{1}{2}} = \tfrac{1}{2}x^{-\frac{1}{2}}dx = \frac{dx}{2\sqrt{x}}.$$

Whence we have the following rule for finding the differential of the square root of a variable quantity; *Divide the differential of the quantity under the radical by twice the radical.*

22. The rules above are sufficient for the differentiation of Algebraic quantities. We will now apply them to some examples.

I. Simple quantities in which the factors are raised to different powers, entire or fractional.

Ex. 1. To find the differential of ax^3y^2. We first consider x^3 and y^2 as two simple variables, which gives, art. 18,

$$d\,(ax^3y^2) = ax^3d.y^2 + ay^2d.x^3,$$

whence, performing the operations,

$$d\,(ax^3y^2) = 2ax^3ydy + 3ay^2x^2dx.$$

Ex. 2. To find the differential of $x^2y^{\frac{5}{2}}$.

Ans. $\frac{5}{2}x^2y^{\frac{3}{2}}dy + 2xy^{\frac{5}{2}}dx.$

Ex. 3. To find the differential of x^3y^{-2}.

Ans. $3x^2y^{-2}dx - 2x^3y^{-3}dy.$

Ex. 4. To find the differential of $x^5y^{-\frac{2}{3}}$.

Ans. $5x^4y^{-\frac{2}{3}}dx - \frac{2}{3}x^5y^{-\frac{5}{3}}dy.$

Ex. 5. To find the differential of $x^{\frac{1}{2}}y^{\frac{1}{2}}$.

Ans. $\dfrac{xdy + ydx}{2x^{\frac{1}{2}}y^{\frac{1}{2}}}.$

II. Quantities composed of simple terms connected by the signs + and —.

Ex. 1. To find the differential of $ax^3 + bx^2 + cxy$. Regarding each term as a simple variable, we have by art. 17,

$$d\,(ax^3 + bx^2 + cxy) = d.ax^3 + d.bx^2 + d.cxy,$$

or performing the operations

$$= 3ax^2dx + 2bxdx + cxdy + cydx.$$

Ex. 2. To find the differential of $ax^2 + bx + \dfrac{cy}{x^2}$.

Ans. $2axdx + bdx - \dfrac{2cydx}{x^3} + \dfrac{cdy}{x^2}.$

Ex. 3. To find the differential of $x^3y - ay^2 + b^2$.

Ans. $3x^2ydx + x^3dy - 2aydy.$

Ex. 4. To find the differential of $ax^{\frac{5}{3}} + \frac{a}{x^2} - bx^{\frac{1}{2}}$.

Ans. $\frac{5}{3}ax^{\frac{2}{3}}dx - \frac{2adx}{x^3} - \frac{bdx}{2\sqrt{x}}$.

Ex. 5. To find the differential of $7x^5 - 4x^2y^3 + \frac{3}{\sqrt{x}}$.

Ans. $35x^4dx - 8xy^3dx - 12x^2y^2dy - \frac{3dx}{2x^{\frac{3}{2}}}$.

III. Compound quantities raised to powers.

Ex. 1. To find the differential of $(a + bx + cx^2)^5$.

Regarding the quantity within the parenthesis as a simple variable, we have by art. 19,

$$\begin{aligned} d(a + bx + cx^2)^5 &= 5(a + bx + cx^2)^4 \times d(a + bx + cx^2) \\ &= 5(a + bx + cx^2)^4 \times (bdx + 2cxdx) \\ &= 5(a + bx + cx^2)^4 (b + 2cx)\, dx. \end{aligned}$$

Ex. 2. To find the differential of $(a + bx^2)^{\frac{5}{3}}$.

Ans. $\frac{10}{3}(a + bx^2)^{\frac{2}{3}}bxdx$.

Ex. 3. To find the differential of $(2ax + x^2)^{\frac{1}{2}}$.

Ans. $\frac{(a + x)\,dx}{(2ax + x^2)^{\frac{1}{2}}}$.

Ex. 4. To find the differential of $\frac{a}{(b^2 + x^2)^3}$.

Ans. $-\frac{6axdx}{(b^2 + x^2)^4}$.

IV. Compound quantities composed of different factors.

Ex. 1. Find the differential of $x^3(a + bx^2)^5$.

In this case we regard each factor as a simple variable and follow the general rule. Thus

$$\begin{aligned} d[x^3(a + bx^2)^5] &= (a + bx^2)^5\, d.x^3 + x^3d(a + bx^2)^5 \\ &= 3x^2(a + bx^2)^5dx + 10bx^4(a + bx^2)^4dx. \end{aligned}$$

Ex. 2. To find the differential of $x^5(a+bx^3)^{\frac{5}{2}}$.

$$\text{Ans.}\quad 5x^4(a+bx^3)^{\frac{5}{2}}dx+\frac{15}{2}bx^7(a+bx^3)^{\frac{3}{2}}dx.$$

Ex. 3. Find the differential of $\dfrac{x^3}{(1+x)^2}$.

$$d.\frac{x^3}{(1+x)^2}=3x^2(1+x)^{-2}dx-2x^3(1+x)^{-3}dx$$

$$=\frac{(3x^2+x^3)\,dx}{(1+x)^3}\quad\text{Ans.}$$

V. Miscellaneous examples.

Ex. 1. Find the differential of $(a^2+x^2)^3$.

Ans. $6\,(a^2+x^2)^2\,xdx$.

Ex. 2. To find the differential of $\dfrac{x}{x+y}$.

Ans. $\dfrac{ydx-xdy}{(x+y)^2}$.

Ex. 3. To find the differential of $(2ax-x^2)^{\frac{1}{2}}$.

Ans. $\dfrac{(a-x)\,dx}{(2ax-x^2)^{\frac{1}{2}}}$.

Ex. 4. To find the differential of $\dfrac{1}{(1-x^2)^{\frac{1}{2}}}$.

Ans. $\dfrac{xdx}{(1-x^2)^{\frac{3}{2}}}$.

Ex. 5. To find the differential of $\dfrac{1+x^2}{1-x^2}$.

Ans. $\dfrac{4xdx}{(1-x^2)^2}$.

Ex. 6. To find the differential of $\dfrac{(b+x)^2}{x}$.

Ans. $\dfrac{(x^2-b^2)\,dx}{x^2}$.

Ex. 7. To find the differential of $\frac{a+2bx}{(a+bx)^2}$.

Ans. $-\frac{2b^2xdx}{(a+bx)^3}$.

Ex. 8. To find the differential of $\frac{x}{(a^2-x^2)^{\frac{1}{2}}}$.

Ans. $\frac{a^2dx}{(a^2-x^2)^{\frac{3}{2}}}$.

Ex. 9. To find the differential of $\frac{x}{(1+x^2)^{\frac{1}{2}}}$.

Ans. $\frac{dx}{(1+x^2)^{\frac{3}{2}}}$.

Ex. 10. To find the differential of $(a^2-x^2)^{\frac{1}{2}}$.

Ans. $-\frac{xdx}{(a^2-x^2)^{\frac{1}{2}}}$.

Ex. 11. To find the differential of $\frac{a}{(a-x)^3}$.

Ans. $\frac{3adx}{(a-x)^4}$.

Ex. 12. To find the differential of $\frac{x+y}{z^3}$.

Ans. $\frac{z(dx+dy)-3(x+y)dz}{z^4}$.

FUNCTIONS.

23. We have, thus far, found the differentials of variables detached from each other, or united without mutual dependence. But the variable quantities which enter into a question are always connected by the conditions of the question; and it is in this relation, or as functions of each other, that, in general, we wish to find their differentials.

A variable quantity is said to be a *function* of another, when they are so connected that a change in the value of the one necessarily produces a change in the value of the other.

Thus, the *height* from which a heavy body has fallen and the *time* of its fall are evidently so connected, that a change in one necessarily involves a change in the other. The time is, therefore, a function of the height; and conversely, the height is a function of the time.

In like manner, in the equation $y = ax + b$, y and x are so connected that they must vary together. They are, therefore, functions of each other. In general, when an equation subsists between several variable quantities, any one of them is a function of all the others.

A quantity, it is evident, is a function of all the quantities, variable or constant, on which its value depends. It is usual, however, to name the variables only.

To indicate simply that a quantity, y for example, depends upon another, x, for its value, we inclose the x in a parenthesis, and place before it some one of the letters f, F. φ, ψ, &c. Thus,

$$y = f(x), \quad y = F(x), \quad y = \varphi(x), \quad \&c.$$

are read, y equal to a function of x.

The letters f, F, φ, &c. when they occur in this connection, are to be regarded simply as signs, indicating that y is found by certain operations upon x and the constants connected with it. Thus if in the equation, $y = f(x)$, we have $f(x) = ax^3$, y is found by raising x to the third power and multiplying the result by a. If $f(x) = \frac{\sqrt{x} + a}{b}$, then y is found by adding a to the square root of x and dividing the result by b.

To indicate that two variables x, y, for example, are mutually dependent, we write

$$f(x, y) = 0, \quad F(x, y) = 0, \quad \&c.$$

which are read function $(x, y) = 0$, &c. If there are three variables, as x, y, z, we write $f(x, y, z) = 0$.

When several variables are functions of each other, we may assign at pleasure values to one or more of them, from which those of the others may be determined. The variables thus selected are called the *independent* variables, the others the *dependent* variables.

Thus in the equation $y = ax + b$, if we assign any arbitrary values to x, those of y will be determined by them. In this case x is the independent, y the dependent variable.

24. Functions, in respect to the manner in which the quantities are connected with each other, are divided into two classes *explicit* and *implicit*.

A function is said to be *explicit* when the operations to be performed upon the variables, in order to obtain the value of the function, are directly expressed; and are, therefore, manifest without the necessity of solving an equation. Thus $y = ax + b$ is an explicit function. For it is obvious, at once, that in order to find y, we must multiply x by a and add b to the product.

An *implicit* function is one in which the operations, necessary to obtain the value of the function, are not directly expressed, but must be determined by solving an equation. Thus y is an implicit function of x in the equation $x^2 + 2xy + a^2 = 0$; for in order to determine the mode of its dependence upon x, we must solve the equation.

25. Functions, also, with respect to the operations which produce them, are divided into two classes, *algebraic* and *transcendental.*

Algebraic functions are those which are formed by the first five processes of Algebra, viz. addition, subtraction, multiplication, division, and the raising to powers entire or fractional.

The elementary forms of these functions are

$$y = a + x,\ y = a - x,\ y = ax,\ y = \frac{a}{x},\ y = x^n,\ y = x^{\frac{1}{n}}.$$

Transcendental functions are those which are not thus formed. They are of different kinds. Their names, with the most simple forms of each, are as follows,

Exponential, as $y = a^x$;

Logarithmic, as $y = \log x$;

Circular, as $y = \sin x$.

26. Functions are also distinguished as *increasing* or *decreasing*.

An increasing function is one which is increased when the variable is increased, or decreased when the variable is decreased. Thus in the function,

$$y = ax + b,$$

if the value of x is increased, that of y will also be increased; if the value of x is diminished, that of y will also be diminished.

A decreasing function is one which is decreased when the variable is increased, and increased when the variable is decreased. Thus in the function,

$$y = (R^2 - x^2)^{\frac{1}{2}},$$

as the value of x is increased, that of y is decreased, and as x decreases the value of y increases.

DIFFERENTIATION OF FUNCTIONS.

27. Let it now be required to find the differential of the function $y = (a + x)^3$. Here, it is evident, we must consider y as becoming $y + dy$, when x becomes $x + dx$. We shall have, therefore, for the differential of the function

$$dy = 3\,(a + x)^2\,dx.$$

By means of this equation a relation, it is evident, is established between the indeterminates dy and dx. Dividing both sides by dx, we have

$$\frac{dy}{dx} = 3\,(a + x)^2.$$

Thus, although dy and dx are both indeterminate, and no absolute values can, therefore, be assigned to them, the ratio of dy to dx has a determinate value expressed by $3(a+x)^2$.

The differentials of variables connected by equations, are found, it is evident, in the same manner, as when they stand alone. The rules which have been given are sufficient, therefore, for the differentiation of all Algebraic functions.

SECTION III.

APPLICATION OF THE DIFFERENTIAL CALCULUS TO THE SOLUTION OF PROBLEMS.

28. We have now seen the nature of the auxiliary quantities employed in the Calculus, their adaptation to their object, and the manner in which they are derived from their primitives. We proceed next to show, by some examples, the aid derived from these auxiliaries in the solution of questions.

PROBLEM OF TANGENTS.

29. This problem may be enunciated as follows.

To determine for each point of a plane curve, the equation of which is given, the direction of its tangent.

This problem early engaged the attention of Mathematicians. A few particular cases are readily solved by common Algebra. But its general solution by aid of the ordinary analysis, is very difficult. The effort to effect this solution was one of the most important as well as immediate steps toward the invention of the Calculus. We will now see how the object is accomplished by aid of this instrument.

30. Let AB be the given curve, AX the axis of x, A the origin, M the given point, and $AP = x$, $PM = y$, its co-ordinates.

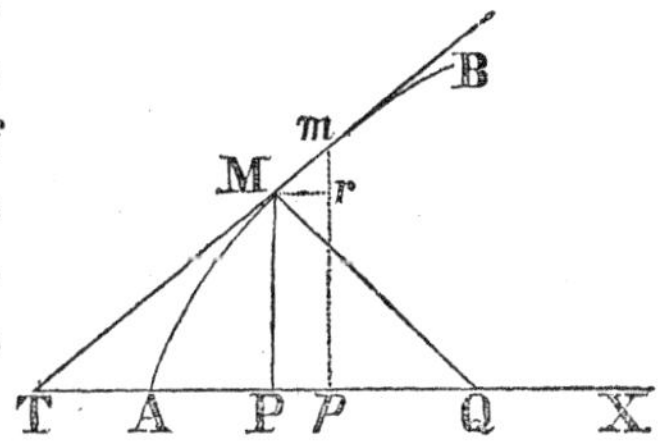

To draw a tangent to the curve AB, we conceive this curve to be a polygon of an infinite number of infinitely small sides. Let Mm be one of these sides. The prolongation of Mm will be a tangent to the curve at the point M. Let Mm be produced to meet the axis of x, also produced, in the point T. To find the direction of the tangent MT, it will be sufficient to determine the value of the subtangent PT.

To find PT, draw the ordinate pm, and draw Mr parallel to AX. Then, since Mm is infinitely small, Mr = Pp, and rm will also be infinitely small; and we shall have M$r = dx$, and $rm = dy$.

But the infinitely small triangle rMm is similar to the finite triangle PTM, and

$$rm : r\mathrm{M} :: \mathrm{PM} : \mathrm{PT},$$

or

$$dy : dx :: y : \mathrm{PT};$$

whence

$$\mathrm{PT} = y\frac{dx}{dy}, \qquad (1)$$

an equation for the subtangent to any point M of the curve.

Let the equation to the curve be

$$y^2 = \frac{\mathrm{B}^2}{\mathrm{A}^2}(\mathrm{A}x - x^2),$$

which is the equation of the ellipse referred to the vertex A.

Differentiating this equation we obtain

$$2ydy = \frac{\mathrm{B}^2}{\mathrm{A}^2}(\mathrm{A} - 2x)\,dx,$$

from which we deduce

$$\frac{dx}{dy} = \frac{2\mathrm{A}^2y}{\mathrm{B}^2(\mathrm{A} - 2x)}. \qquad (2)$$

Substituting this value in equation (1), we obtain

$$\mathrm{PT} = \frac{2\mathrm{A}^2y^2}{\mathrm{B}^2(\mathrm{A} - 2x)}.$$

Thus we have obtained an expression for the subtangent to the ellipse in terms of the co-ordinates x and y, and the constants which enter into its equation. We may now eliminate one of these co-ordinates, y for example, by means of the equation to the curve, which gives finally

$$PT = \frac{Ax - x^2}{\frac{1}{2}A - x}, \qquad (3)$$

a result precisely the same with that at which we arrived, though in a manner far less expeditious, by common Algebra.

If the preceding solution be examined with attention, it will be seen

1°. That an equation for the problem has been established, with great facility, by aid of the infinitely small quantities dx and dy employed as auxiliaries.

2°. These auxiliaries are derived from the quantities which primarily come under consideration in the problem; viz. the co-ordinates x and y of the point on the curve through which the tangent is required to be drawn.

3°. By very simple rules deduced from their relation to their primitives, we have been able, from the equation to the curve, to find the ratio, dx to dy, of these auxiliaries, in terms of their primitives and the constants which enter into the equation.

4°. By means of the ratio thus found, we have been enabled, after they have served their purpose, to eliminate the indeterminates dx, dy, or the auxiliaries employed, and the result is left, as it should be, in finite and determinate terms.

31. The expression for the subtangent PT, equation (1), applies, it is evident, to all plane curves referred to rectangular co-ordinates. Hence, for a particular curve, we have only to find the value of the ratio dx to dy, substitute and reduce, as in the preceding example. The value of this ratio, it is evident, can always be found when the equation to the curve is given. For if we differentiate the equation there will result, it is easy to see, two kinds of terms only, viz. 1°. terms multiplied by dx; 2°. terms multiplied by dy. We shall

be able, therefore, in every particular case, to eliminate the indeterminates employed, and obtain the subtangent, finally, in terms of x, y, and constants only. The problem is, therefore, completely solved.

Ex. 1. Find the subtangent to the common parabola, the equation of which is $y^2 = 2px$. Ans. $2x$.

Ex. 2. Find the subtangent to the ellipse, when the origin is at the centre, and the equation $A^2y^2 + B^2x^2 = A^2B^2$.

Ans. $\dfrac{x^2 - A^2}{x}$.

Ex. 3. Find the subtangent to the circle, its equation when it is referred to the centre being $y^2 + x^2 = R^2$.

Ans. $\dfrac{x^2 - R^2}{x}$.

Ex. 4. Find the subtangent to the curve $xy = a^2$.

Ans. $-x$.

32. In the solution of the problem of tangents, above, we have a simple illustration of the peculiar feature of the Calculus, its distinction from common Algebra, and its power as an instrument of investigation.

1°. That which belongs exclusively to the Calculus in the solution is the employment, it is evident, of certain infinitely small quantities as auxiliaries to facilitate the establishment of an equation for the problem; and the use of special rules, derived from the relation of these auxiliaries to their primitives and designed to effect their final elimination after they have accomplished the object for which they are introduced. All the other processes, or operations, are merely those of common Algebra.

2°. The power of the Calculus is seen in the remarkable facility with which the problem is solved; and, especially, in the extreme generality of the solution,—one simple and easily obtained differential expression being sufficient to give the tangents to all plane curves, referred to rectangular axes.

33. Since the tangent to any point of a curve makes a right angle with the normal, the direction of the tangent will, also, be known when that of the normal is found.

Let it be required, then, as a second solution of the problem proposed, to find the subnormal PQ to any point M of a plane curve AM.

The same notation being employed as before, we have from the similar triangles Mrm, MPQ,

$$dx : dy :: y : \text{PQ};$$

whence $$\text{PQ} = y\frac{dy}{dx},$$

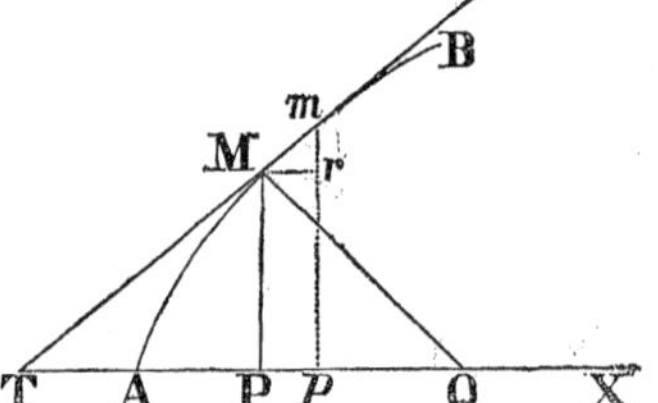

a differential equation for the subnormal PQ common, it is evident, to all plane curves. Having established this equation, the problem is now reduced, as in the case above, to a simple problem of analysis, viz.; the elimination of the infinitesimals dx and dy employed as auxiliaries. This, as we have seen in the preceding solution, may always be done by very simple and uniform methods. For any particular curve we have only to differentiate the equation to the curve, deduce the value of the ratio of dy to dx, substitute, and reduce as in the preceding examples.

Ex. 1. To find the subnormal to the ellipse, the equation to the curve being $A^2y^2 = B^2(Ax - x^2)$.

Ans. $\text{PQ} = \frac{B^2}{A^2}(\frac{1}{2}A - x)$.

Ex. 2. Find the subnormal to the parabola, whose equation is $y^2 = 2px$. Ans. p.

Ex. 3. Find the subnormal to the circle, the equation being $y^2 + x^2 = R^2$. Ans. $-x$.

Ex. 4. Find the subnormal to the hyperbola whose equation is $y^2 = 2(ax + x^2)$. Ans. $a + 2x$.

34. In the triangle mrM (fig. art. 33) we have, radius being 1,

$$1 : \mathrm{T} :: \mathrm{M}r : rm :: dx : dy;$$

whence
$$\mathrm{T} = \frac{dy}{dx},$$

in which T is the trigonometrical tangent of the angle $m\mathrm{M}r$, or which is the same thing, the angle MTP, which the tangent line MT makes with the axis of x. The value of T being found in finite terms, by means of the equation to the curve, the direction of the tangent will be known. We thus have a third solution of the problem proposed.

Ex. Let the equation to the curve be $y^2 = 4x$, to find the direction of the tangent.

$$\mathrm{T} = \frac{1}{\sqrt{x}}. \quad \text{Ans.}$$

To draw a tangent to any point on a curve whose equation is given, it will, in general, be found most convenient to employ the expression for the subtangent PT. We have found the expressions for the subnormal PQ, and the trigonometrical tangent T, chiefly as an additional exercise for the learner upon the problem of tangents.

MAXIMA AND MINIMA.

35. The greatest value of a variable quantity is called a *maximum*, the least a *minimum*.

Problems concerning maxima and minima are among the most important that occur in analysis, and in the practical applications of mathematics. In further illustration of the Calculus we proceed to some problems of this description.

Ex. 1. Let it be required to divide a number a into two parts such that their product may be the greatest possible.

Let $x =$ one of the parts, then $a - x =$ the other, and we have for the product

$$x(a - x) = ax - x^2;$$

but, by hypothesis, this product is now the greatest possible; its increment, or differential must, therefore, be zero, and we have

$$d\,(ax - x^2) = 0\,;$$

or performing the differentiation, and reducing,

$$x = \tfrac{1}{2}a.$$

The product, therefore, is the greatest possible, when the number is divided into two equal parts, as we have seen already in Algebra.

Ex. 2. To divide a number a into two parts such that the sum of the squares of the parts shall be the least possible.

Let $x =$ one of the parts, then $a - x =$ the other, and by the question

$$x^2 + (a - x)^2 = \text{a minimum}\,;$$

the quantity being the least possible, its differential will be 0, and we have

$$d\,[x^2 + (a - x)^2] = 0\,;$$

whence, differentiating and reducing,

$$x = \tfrac{1}{2}a.$$

The number, therefore, must be divided into two equal parts.

Ex. 3. It is required to find a fraction such that the fraction shall exceed its cube by the greatest possible quantity.

Let x represent the fraction; then by the question

$$x - x^3 = \text{a maximum}\,;$$

and

$$d\,(x - x^3) = 0.$$

Ans. The fraction is the square root of one third.

36. The preceding method consists, it is obvious, in finding from the conditions of the problem the expression required to be a maximum or minimum, and then putting the differential of this expression equal to 0. It will be easy to see how we are to proceed in similar cases. In the following examples we employ two unknown quantities, in order to obtain with more facility the expression required to be a maximum or minimum.

Ex. 1. Among right triangles of the same hypothenuse to determine that which has the greatest area.

Let $a =$ the given hypothenuse, x one of the sides and y the other; then

$$x^2 + y^2 = a^2; \qquad (1)$$

and by the question

$$\tfrac{1}{2}xy = \text{a maximum}. \qquad (2)$$

Differentiating this last, and reducing

$$xdy + ydx = 0; \qquad (3)$$

an equation involving both dx and dy. Differentiating, therefore, the first equation we obtain

$$xdx + ydy = 0; \qquad (4)$$

eliminating dx, for example, from (3) by means of (4), we obtain

$$x^2dy - y^2dy = 0;$$

whence

$$x = y.$$

Ans. The two remaining sides must be equal.

Ex. 2. A tin-smith was ordered to make a cylindrical vessel, which should hold a given quantity, a gallon for example, and be the lightest possible; required the proportion between the radius of its base and height.

Let c^2 be the required capacity of the vessel, x its altitude, y the radius of the base; then $\pi y^2x = c^2$; and the surface $= \pi y^2 + 2\pi yx$. And since the tin-plate is of uniform thickness we shall have

$$\pi y^2 + 2\pi yx = \text{a minimum}.$$

Ans. The height of the cylinder must be equal to the radius of its base.

Ex. 3. Among parallelopipids of the same surface and the same altitude, to determine that whose solidity is the greatest.

Let c^2 be the given surface, h the given altitude, and x and y, respectively, the sides of the base; then

$$2hx + 2hy + 2xy = c^2;$$

and

$$hxy = \text{a maximum}.$$

Ans. That in which the base is a square.

Ex. 4. A carpenter has a quantity, a feet, of boards which he wishes to make into a box, that shall be of the greatest possible capacity. What must be the form of the box?

By the preceding problem it is evident that the base of the box must be a square; and it remains only to determine the relation of the altitude to a side of the base.

Ans. The box must be a cube.

37. From what has been done the following miscellaneous examples will now be readily solved.

Ex. 1. To divide a number $2a$ into two parts such that the sum of the square roots of the parts shall be a maximum.

Ans. The number must be divided into two equal parts.

Ex. 2. To divide a given number a into two factors the sum of which shall be a maximum.

Ans. The two factors must be equal.

Ex. 3. The difference between two numbers is a; required that the square of the greater divided by the less shall be a minimum.

Ans. The greater is twice the less.

The preceding examples have been solved by the processes of Algebra exclusively, Algb. art. 119. Let the learner now compare the two methods.

Ex. 5. A gentleman purchased a garden plot to be set off to him in the form of a rectangle, with the single condition that the sum of two contiguous sides must be equal to 50 rods. What dimensions must he take for the sides in order to secure the greatest amount of land for his bargain.

Ans. Each must be equal to 25 rods.

Ex. 6. A gentleman has a park in the form of a triangle, the base of which is a feet, and the perpendicular b feet. In this he wishes to make the greatest rectangular garden possible, one of the sides of which is to be on the base; it is required to find how many feet from the vertex the other side must be drawn.

Ans. $\frac{b}{2}$ feet; or the perpendicular must be bisected.

We shall return to a more full discussion of the question of maxima and minima, as well as to that of the problem of tangents.

....

SECTION IV.

INTEGRAL CALCULUS.

38. In the preceding problems, the quantities sought enter *directly* into the differential equations formed to aid in their solution. We have, in consequence, been able to eliminate the auxiliaries employed, by the processes of ordinary Algebra alone. But there are numerous problems, and among them some of the most important that can be presented, in which the quantities sought do not enter thus directly into the differential equations obtained, but only *indirectly* through their differentials. When this is the case, our only resource for the complete elimination of all the auxiliaries employed, is to go back, by a reverse process, from the differential equations finally obtained to the primitive or finite equations from which the differential equations must be regarded as derived.

Let s be the quantity sought, and let the differential equations first obtained be reduced ultimately to an equation of the form

$$ds = f(x)\,dx;$$

here, the only method by which we can eliminate the auxiliaries employed, is to reverse the process by which this equation must be regarded as derived from the finite equation which it is our object to establish.

We have already solved for simple algebraic quantities the general question,—*Given a variable quantity to find its element or differential.* We are now led to a second general question the reverse of this, viz.—*Given the element or differential of a variable quantity to find the quantity itself.*

39. Since a variable quantity is supposed to arrive at a determinate magnitude by the successive addition of its differential or ele-

ment, the value of the variable, at any particular stage of its increase, will be equal, it its evident, to the sum of all the elementary portions which are supposed to be united in its formation. The process, therefore, by which we determine the value of a variable quantity from its element, or which is the same thing, by which we return from the auxiliary to its primitive, is called *Integration*— a summing up. The result obtained is called the sum, or *Integral ;* and that part of the Calculus which relates to the finding of integrals is called the INTEGRAL CALCULUS.

INTEGRATION. SIMPLE QUANTITIES.

40. An integral is indicated by placing the letter $\int$, signifying *sum of* or *integral*, before the differential expressions. Thus, $\int x^2dx$ indicates the integral of x^2dx.

1. To find the integral of a differential quantity we reverse, it is obvious, the process of differentiating. Thus, since we find the differential dx of x by writing the letter d before the x; conversely, we find the integral of dx by removing the d from before it.

2. To find the integral of $3x^2dx$. By art. 19,

$$d.\, x^3 = 3x^{3-1}dx = 3x^2dx.$$

Conversely, therefore, $\int 3x^2dx = \frac{3x^{2+1}dx}{3dx} = x^3.$

In like manner $\qquad d.x^m = mx^{m-1}dx.$

Conversely, therefore,

$$\int mx^{m-1}dx = \frac{mx^{m-1+1}dx}{mdx} = x^m.$$

To find, therefore, the integral of a monomial differential of the form $mx^{m-1}dx$, we have the following rule,

Increase the exponent of the variable by unity, divide by the exponent thus increased and by the differential of the variable.

41. We have seen, (art. 17), that if a variable quantity is affected with a constant factor, the differential will be affected with the

same factor. The converse will be the case, therefore, in the process of integration. Thus,

$$\int abx^2dx = ab\int x^2dx = \tfrac{1}{3}abx^3.$$

If then the quantity to be integrated is affected with a constant factor, *this factor may be placed without the sign of integration.*

It has been seen, moreover, that a constant quantity connected with a variable by the sign $+$ or $-$, disappears in the process of differentiation. Thus the same differential may answer to several integrals, which differ from each other only in the constant term. In the process of integration, therefore, a constant must be added to the result obtained. Thus, the integral of $3ax^2dx$ should be written

$$3a\int x^2dx = ax^3 + \textit{constant};$$

or

$$3a\int x^2dx = ax^3 + C,$$

C being the constant necessary to complete the integral. The constant is entirely arbitrary so long as it is undetermined, and is, therefore, called the *arbitrary* constant. But when the Calculus is applied to the solution of a real problem, the conditions of the problem will enable us always to determine the constant. When the constant is thus determined, the integral is said to be *corrected.*

42. Let it be required, for example, to integrate the function $dy = adx$, the differential equation of the right line. From what has been said we have for the general integral

$$y = ax + C.$$

If now the right line is required to pass through the origin, for the value $x = 0$, we have $y = 0$, which gives $C = 0$ also. We have, therefore, for the *corrected* integral

$$y = ax.$$

If the right line is required to intersect the axis of y at the distance b from the origin, then for $x = 0$, $y = b$, and the corrected integral is

$$y = ax + b.$$

The value of the constant must, it is evident, be so determined as to be true for every value which can be attributed to the variables. In general, therefore, if the constant be correctly determined for any one value of the variables, it will be determined for all.

In the examples which follow we omit the constant. It will be recollected, however, that a constant must always be added to the integrals obtained.

43. Let it be required next to integrate the quantity $\frac{dx}{x^3}$. This returns, it is evident, to $x^{-3}dx$; whence

$$\int\frac{dx}{x^3}=\int x^{-3}dx=-\tfrac{1}{2}x^{-2}=-\frac{1}{2x^2}.$$

In like manner
$$\int\frac{adx}{x^{m+1}}=-\frac{a}{mx^m}.$$

The rule fails in the case where we have $\frac{dx}{x}$ to integrate. We shall show hereafter the value of this particular integral.

To find next the integral of $ax^{\frac{2}{3}}dx$. We have

$$a\int x^{\frac{2}{3}}dx=\frac{ax^{\frac{2}{3}+1}}{\frac{2}{3}+1}=\tfrac{3}{5}ax^{\frac{5}{3}}.$$

In like manner

$$\int\frac{dx}{x^{\frac{1}{2}}}=\int x^{-\frac{1}{2}}dx=2x^{\frac{1}{2}}.$$

Thus, if the differential of a variable be divided by the square root of the variable, *the integral will be twice the square root of the variable.*

COMPLEX QUANTITIES.

44. The differential of a quantity composed of several terms is equal, we have seen (art. 17), to the differential of the several terms taken each with its proper sign; conversely, therefore, the integral of a differential expression composed of several differential

terms *will be equal to the sum of the integrals of the separate terms, taken each with its proper sign.*

Ex. 1. Let it be required to find the integral of the function

$du = adx - \frac{bdx}{x^3} + x^{\frac{3}{2}}dx,$

$$\int du = \int adx - \int \frac{bdx}{x^3} + \int x^{\frac{3}{2}}dx\,;$$

whence $$u = ax + \frac{b}{2x^2} + \tfrac{2}{5}x^{\frac{5}{2}}.$$

Ex. 2. To find the integral of the function

$$du = ax^2dx + x^{\frac{1}{2}}dx + \frac{dx}{x^5}.$$

Ans. $u = \frac{1}{3}ax^3 + \frac{2}{3}x^{\frac{3}{2}} - \frac{1}{4x^4}.$

45. Let it be required next to find the integral of the function $du = (bx + cx^2)^2dx.$

Expanding the quantity within the parenthesis and multiplying by dx we obtain

$$du = b^2x^2dx + 2bcx^3dx + c^2x^4dx\,;$$

whence $$u = \tfrac{1}{3}b^2x^3 + \tfrac{1}{2}bcx^4 + \tfrac{1}{5}c^2x^5.$$

In general, if a polynomial be of the form $(a + bx + cx^2..)^n dx$, n being a positive whole number, to find the integral, *we raise the quantity within the parenthesis to the given power, multiply each term by* dx, *and then integrate each term separately.*

Ex. 1. Find the integral of $(a + bx)^2dx.$

Ans. $a^2x + abx^2 + \frac{1}{3}b^2x^3.$

Ex. 2. Find the integral of $(1 + 2x + 3x^2)^2dx.$

Ans. $x + 2x^2 + \frac{10}{3}x^3 + 3x^4 + \frac{9}{5}x^5.$

46. We have seen, (art 22), that in order to find the differential of the power of a polynomial, we multiply by the exponent, diminish the exponent by unity, and then multiply by the differential of the polynomial. Conversely, therefore, if in a polynomial dif-

ferential, the quantity outside the parenthesis is the differential of that within; to find the integral, *Increase the exponent of the polynomial by unity, divide by the exponent thus increased and by the differential of the polynomial.*

Ex. 1. Thus to find the integral of $3(ax+x^2)^2(adx+2xdx)$ we have

$$\int 3(ax+x^2)^2(adx+2xdx) = \frac{3(ax+x^2)^{2+1}(adx+2xdx)}{3(adx+2xdx)}$$
$$= (ax+x^2)^3.$$

Ex. 2. To find the integral of $(a+3x^2)^3 6xdx$.

Ans. $\frac{1}{4}(a+3x^2)^4$.

Ex. 3. To find the integral of $(a+x^2+x^3)^2(2x+3x^2)dx$.

Ans. $\frac{1}{3}(a+x^2+x^3)^3$.

2. The rule is equally applicable, if the quantity without the parenthesis is the differential of that within, multiplied or divided by a constant quantity.

Thus, $\int(a+3x^2)^2 xdx = \frac{1}{18}(a+3x^2)^3$.

Ex. 1. Find the integral of $(a+bx^2)^3 xdx$.

Ans. $\dfrac{(a+bx^2)^4}{8b}$.

Ex. 2. Find the integral of $(a+\frac{1}{2}x)^4 dx$.

Ans. $\frac{2}{5}(a+\frac{1}{2}x)^5$.

Ex. 3. Find the integral of $(a^2-x^2)^5 xdx$.

Ans. $-\frac{1}{12}(a^2-x^2)^6$.

Ex. 4. Find the integral of $2(a+bx-cx^2)^3(b-2cx)dx$.

Ans. $\frac{1}{2}(a+bx-cx^2)^4$.

3. The rule also applies equally when the exponent of the polynomial is fractional, positive or negative.

Ex. 1. To find the integral of $\dfrac{(a^2+2ax)dx}{(ax+x^2)^{\frac{1}{2}}}$.

$$\int\frac{(a^2+2ax)dx}{(ax+x^2)^{\frac{1}{2}}} = \int a(ax+x^2)^{-\frac{1}{2}}(a+2x)dx$$
$$= 2a\,(ax+x^2)^{\frac{1}{2}}.$$

Ex. 2. Find the integral of $(a^2+x^2)^{\frac{1}{2}}xdx$.

Ans. $\frac{1}{3}(a^2+x^2)^{\frac{3}{2}}$.

Ex. 3. Find the integral of $\dfrac{xdx}{(a^2+x^2)^{\frac{1}{2}}}$.

Ans. $(a^2+x^2)^{\frac{1}{2}}$.

Ex. 4. Find the integral of $3(a^4-x^4)^{\frac{5}{3}}x^3dx$.

Ans. $-\frac{9}{32}(a^4-x^4)^{\frac{8}{3}}$.

Ex. 5. Find the integral of $(a+bx^2)^{\frac{1}{2}}mxdx$.

Ans. $\dfrac{m}{3b}(a+bx^2)^{\frac{3}{2}}$.

Ex. 6. Find the integral of $(a+bx^n)^m x^{n-1}dx$.

Ans. $\dfrac{(a+bx^n)^{m+1}}{(m+1)nb}$.

The preceding rules are sufficient for the integration of the more simple Algebraic functions. We shall now apply them to the solution of some problems.

SECTION V.

APPLICATION OF THE INTEGRAL CALCULUS TO THE SOLUTION OF PROBLEMS.

47. We are now prepared for the complete solution of the problem with which we commenced (art. 2), viz.

To find the space through which a body, acted upon by gravity, will descend in a determinate time.

Putting dt for the infinitely small element of the time, and ds for the corresponding element of the space, we easily obtain, in the manner already explained, an equation for the problem, viz.

$$ds = gtdt, \quad (1)$$

in which s, the quantity sought, enters only by its differential ds which with dt is employed as auxiliary to the establishment of the equation. To eliminate the auxiliaries ds and dt, we must now go back, by a reverse process, from the differentials of which the equation is composed, to the finite quantities from which they are derived and the relation of which we wish to determine. In other words, we must apply to this equation the process of integration, for which we have obtained the necessary rules. Performing the operation we have

$$s = \tfrac{1}{2}gt^2 + C. \qquad (2)$$

To determine the constant C, it will be observed that when $t = 0$, we shall have, also, $s = 0$. That is, there will be no constant to be added; and we shall have finally

$$s = \tfrac{1}{2}gt^2. \qquad (3)$$

Thus the space through which a body, acted upon by gravity, will descend in a given time is equal to the square of the time multiplied by $\frac{1}{2}g$, g being the velocity acquired by the body in the first instant of time, and found by experiment to be 32.2 feet nearly.

It will be easy to see the part performed by the Calculus in the solution of this problem. We pass next to some problems relative to the determination of areas, the volume and surfaces of solids, and the rectification of curve lines.

QUADRATURE OF CURVES.

48. The *quadrature* of a curve consists in finding for it an equivalent rectilineal area. When this can be done in a limited number of finite terms, the curve is said to be *quadrable*, and may be represented by an equivalent square.

PROB. *To find the area of a plane curve referred to rectangular co-ordinates.*

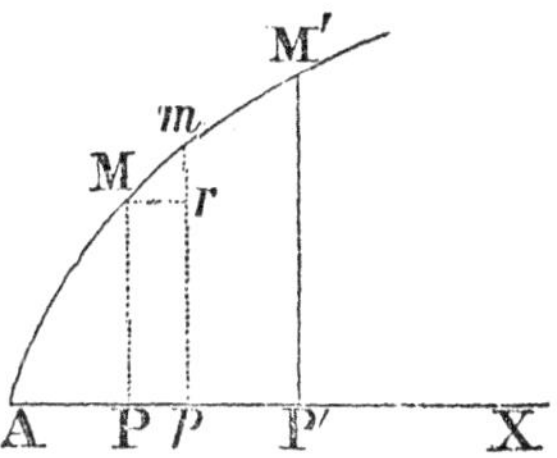

Let AP′M′ be the curve surface whose area is sought. Let PM be any ordinate of the curve, and AP the corresponding abscissa. Let Pp be the infinitely small increment or differential of the abscissa AP. Draw the ordinate pm. Then PpmM will be the infinitely small increment or differential of the surface APM, corresponding to the increment Pp of the abscissa AP. But Pp being infinitely small, Mm will, also, be infinitely small. And the curve being considered a polygon of an infinite number of infinitely small sides, Mm may be regarded as one of these sides, and the figure PpmM will be a trapezoid. Putting A for the area sought, $\mathrm{AP} = x$, $\mathrm{PM} = y$, we shall have

$$d\mathrm{A} = \frac{\mathrm{PM} + pm}{2} \times \mathrm{P}p = \frac{2y + dy}{2} \times dx = ydx + \frac{dydx}{2}.$$

But $dydx$ is an infinitely small quantity of the second order compared with ydx, and should, therefore, be struck out. We shall have, then, for the equation of the problem,

$$d\mathrm{A} = ydx. \qquad (1)$$

The equation to the curve being given, we shall next be able to find the value of y in terms of x, or the value of dx in terms of y and dy. Substituting either of these values, that of y for example, in equation (1), we may then, by the process of integration, eliminate the auxiliaries employed, and the problem will be completely solved.

Ex. 1. To find the area of the common parabola, the equation to which, p being the parameter, is $y^2 = px$.

Deducing the value of y from this equation, we have

$$y = p^{\frac{1}{2}} x^{\frac{1}{2}};$$

whence by substitution in the expression for $d\text{A}$,

$$d\text{A}=p^{\frac{1}{2}}x^{\frac{1}{2}}dx; \qquad (2)$$

taking the integral we obtain

$$\text{A}=\frac{2}{3}p^{\frac{1}{2}}x^{\frac{3}{2}}+C. \quad (3)$$

To determine next the constant, if we put $x=0$, the surface A will be 0; whence the constant $C=0$, and we have for the *corrected* integral

$$\text{A}=\frac{2}{3}p^{\frac{1}{2}}x^{\frac{3}{2}}; \qquad (4)$$

which gives the surface reckoned from the point A, or from the vertex of the parabola.

49. Let it now be supposed, which is the most frequent case, that we wish to find the value of the surface between certain limits, as the area PM M′P′ comprised between the ordinates PM, P′M′, for which we have, as corresponding values of x, $x=a$, $x=b$. Substituting a and b successively in the indefinite integral (3) and subtracting the former of the values thus obtained from the latter, we obtain, still retaining A to indicate the area sought,

$$\text{A}=\frac{2}{3}p^{\frac{1}{2}}b^{\frac{3}{2}}+C-\frac{2}{3}p^{\frac{1}{2}}a^{\frac{3}{2}}-C=\frac{2}{3}p^{\frac{1}{2}}(b^{\frac{3}{2}}-a^{\frac{3}{2}}). \quad (5)$$

When the value of an integral is taken between two successive values a and b of the principal variable, the original expression is said to be integrated between the limits $x=a$, $x=b$; and the integral, thus obtained, is called a *definite integral.*

Thus the value of A, obtained in equation (3), is an *indefinite* integral; that obtained in (4) a *corrected* integral, and that obtained in (5) a *definite* integral.

In the process of finding the definite integral the constant is necessarily eliminated by the subtraction. In general, if we wish to make the constant disappear, *we give two successive values to the independent variable, and take the difference between the two integrals corresponding to these values.*

To indicate integration between limits, or that the definite integral is required between the limits $x = a$, $x = b$, for example, we write the letters a and b with the sign of integration thus, $\int_a^b$. If, for example, the integral of the quantity nx^2dx is required between the limits $x = a$, $x = b$, we have

$$\int_a^b nx^2dx = \tfrac{1}{3}nb^3 - \tfrac{1}{3}na^3 = \tfrac{1}{3}n(b^3 - a^3).$$

50. Let us take next the general equation $a^{n-1}y = x^n$, which represents the whole family of parabolas.

We obtain in the same manner as above, the area being reckoned from the vertex,

$$A = \frac{1}{n+1} \cdot \frac{x^{n+1}}{a^{n-1}},$$

or, substituting y for its value from the equation,

$$A = \frac{1}{n+1} xy.$$

From this result, it will be seen that all parabolic areas are *quadrable*, that is, they can be expressed in terms of the co-ordinates x and y. They are also a determinate part of the circumscribed parallelogram expressed by the fraction $\frac{1}{n+1}$. Thus, in the common parabola $n = \frac{1}{2}$; whence $A = \frac{2}{3}xy$; or the area, as we have before seen, *is equal to two-thirds of the circumscribed parallelogram.*

If $n = 1$, the parabola becomes a right line, and $A = \frac{1}{2}xy$, the area of a triangle in which x is the base and y the altitude.

51. The quadrature of the parabola was discovered by Archimedes, and his solution is memorable as being the first complete one for the quadrature of any curve. A comparison of the methods pursued by him to attain this object with the solution we have just obtained, will strikingly exhibit the aid we derive from the Calculus. We proceed to some additional examples of quadratures.

Ex. 1. To find the area of the curve represented by the equation $y = x - x^3$.

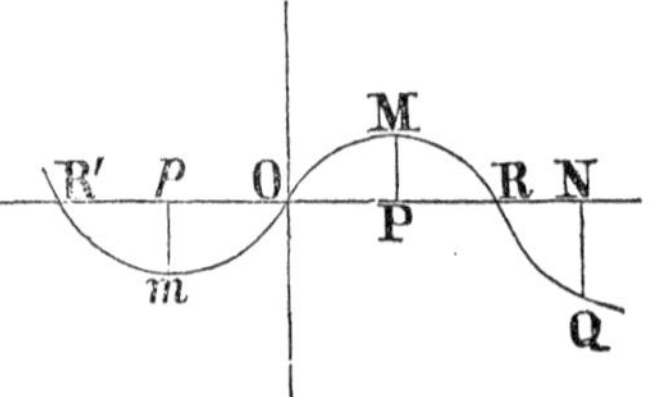

The curve is of the form represented in the figure. It passes through the origin, and intersects the axis of x, at the points $x = 1$, $x = -1$.

Integrating between the limits $x = 0$, $x = 1$, we obtain $\frac{1}{4}$ for the area OMR. Integrating between $x = 0$, $x = -1$, we obtain $0m$R′ equal also to $\frac{1}{4}$. Thus the whole area between R and R′, or the limits $x = 1$, $x = -1$, is $\frac{1}{2}$.

Ex. 2. Find the area of the curve $y = x^3 - b^2x$.

The curve is of the same general form with the preceding. And its area between the limits $x = 0$, $x = b$, is $\frac{1}{4}b^4$. Between $x = b$, $x = -b$, it is $\frac{1}{2}b^4$.

Ex. 3. To find the area of the curve $y = x^3 + ax^2$.

The curve meets the axis at the origin, and intersects it at $x = -a$. Its area between $x = 0$, $x = -a$, is $\frac{1}{12}a^4$. Between $x = 0$, $x = +a$, it is $\frac{7}{12}a^4$.

Ex. 4. To find the area of the curve $ay = x(a^2 - x^2)^{\frac{1}{2}}$.

The curve is called the *knot*. Its general form is that of the figure eight. It intersects the axis at the origin, and at $x = a$, and $x = -a$. It consists of four branches and the whole area is $\frac{4}{3}a^2$.

Ex. 5. To find the area of the curve $y = x(x^2 - a^2)^3$.

Ans. $\frac{1}{8}(x^2 - a^2)^4 + C$.

Let the learner construct the curve geometrically and find its area between the limits $x = 0$, $x = a$.

Ex. 6. Find the area of the curve $xy^2 = a^3$, between the limits $y = b$, $y = c$.

Ans. $2a^3\dfrac{b - c}{bc}$.

Ex. 7. Find the area of the curve $y = x^3 - bx^2$, between the limits $x = 0$, $x = b$. Ans. $\frac{1}{12}b^4$.

Ex. 8. Find the area of the curve $y = 5 + (x - 3)^{\frac{1}{2}}$ between the limits $x = 3$, $x = 12$. Ans. 63.

Ex. 9. Find the area of the curve $y = ax^2$.

Ans. $\frac{1}{3}xy$.

II. CUBATURE OF SOLIDS.

52. The *cubature* of a solid is the finding its solid contents, or the *cube* to which it is equal.

To find the volume or solid contents of a solid of any form, we consider the solid as made up of an infinite number of infinitely thin parallel segments, any one of which may be regarded as its element or differential.

Prob. 1. *To find the volume of a solid of revolution.*

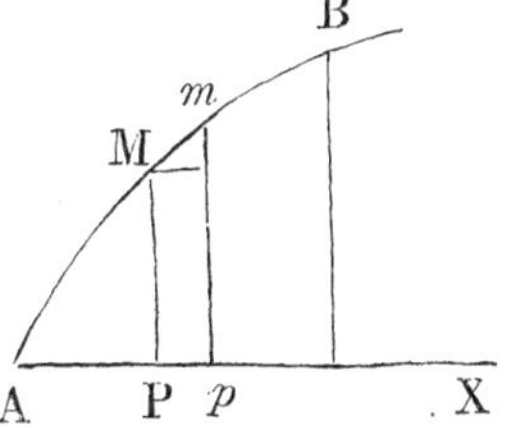

Let the solid be generated by the revolution of the curve AB about AX, the axis of x, taken as the axis of revolution. Let AM be any arc of the generating curve, $AP = x$, $PM = y$.

The area of the circle which has PM for its radius is πy^2; and if this area be multiplied by $dx = Pp$, we shall have the volume of an infinitely thin segment of the solid, which will be its element or differential. Putting V = the whole volume of the solid, we have

$$dV = \pi y^2 dx, \qquad (1)$$

which will be the differential equation for the problem when the curve is revolved about the axis of x. If the curve is revolved about the axis of y, we obtain in like manner for the equation

$$dV = \pi x^2 dy. \qquad (2)$$

It will be easy to see how we are to proceed to the complete solution of the problem in any particular case which may present itself.

Ex. 1. As a first example, let it be required to find the solidity of a prolate spheroid, or that formed by the revolution of an ellipse about its major axis.

From the equation of the ellipse referred to the centre, viz. $A^2y^2 + B^2x^2 = A^2B^2$, we obtain

$$y^2 = \frac{B^2}{A^2}(A^2 - x^2);$$

substituting in the general equation (1), we obtain for the ellipsoid

$$dV = \pi\frac{B^2}{A^2}(A^2 - x^2)\,dx;$$

whence by integration and reduction

$$V = \tfrac{1}{3}\pi\frac{B^2}{A^2}(3A^2x - x^3) + C.$$

Estimating the solidity from the plane passing through the centre, we have for $x = 0$, $V = 0$, and consequently $C = 0$; whence, taking the integral between the limits $x = 0$, $x = A$, we obtain

$$V = \tfrac{2}{3}\pi B^2 \times A;$$

whence, doubling to obtain the whole ellipsoid,

$$2V = \tfrac{2}{3}\pi B^2 \times 2A.$$

But πB^2 expresses the area of the circle described on the conjugate axis, and $2A$ is the transverse axis; hence *the solidity is equal to two-thirds of the circumscribed cylinder.*

If we suppose $A = B$, we shall obtain the solidity of the sphere $= \tfrac{2}{3}\pi R^2 \times 2R$, R being the radius of the sphere.

Ex. 2. To find the solidity of an oblate spheroid, or that generated by the revolution of an ellipse about the minor axis.

Substituting, in the general equation (2), the value of x^2 derived from the equation of the ellipse, we have for the required equation,

$$dV = \pi\frac{A^2}{B^2}(B^2 - y^2)dy;$$

and we find as above

$$2V = \frac{2}{3}\pi A^2 \times 2B,$$

or, the solidity equal to two-thirds the circumscribed cylinder.

Ex. 3. To find the solidity of a paraboloid, the equation to the parabola being $y^2 = 2px$.

Proceeding as above we obtain

$$V = \pi px^2 + C.$$

Reckoning from the vertex the constant will be 0. And if we integrate between $x = 0$, $x = h$, we obtain

$$V = \pi ph^2.$$

Putting $b =$ the ordinate corresponding to h, and substituting for p its value, derived from the equation $b^2 = 2ph$, we obtain

$$V = \pi b^2 \times \frac{h}{2};$$

that is, *the solidity is equal to half the cylinder of the same base and altitude.*

Ex. 4. To find the solidity of a right cone with a circular base, the equation to the generating triangle being $y = ax$.

Ans. Area of the base into one-third the altitude.

Ex. 5. To find the volume of the solid generated by the revolution of the cubical parabola, $y^3 = cx$, about its axis.

Ans. Three-fifths of the circumscribed cylinder.

53. From what has been done, it will be easy to see how we are to proceed in other cases.

Prob. 2. *To find the volume of a solid other than that of revolution.*

Ex. 1. Let it be required to find the solidity of a triangular pyramid.

Let b^2 represent the area of the base, h the altitude of the pyramid, and x any distance from the vertex. The area of the section at the distance x will be $\frac{b^2x^2}{h^2}$, and we shall have

$$dV = \frac{b^2x^2}{h^2}dx,$$

from which we find the whole solidity equal to the area of the base by a third of the altitude.

Ex. 2. To find the solidity of the groin.

The groin is a solid generated by a variable rectangle, the plane of which moves always parallel to itself.

Let the variable rectangle be a square; and let the two co-ordinate sections perpendicular to the base through the middle of the opposite sides of the square be equal semicircles. Let A be the vertex of the groin, x the distance, from the vertex, of any section parallel to the base, and $y^2 = 2ax - x^2$ the equation of the circle referred to A as the origin. We shall find for the element

$$dV = 4(2ax - x^2)\,dx.$$

Taking the integral between $x = 0$, $x = a$, we obtain

$$V = \frac{8}{3}a^3.$$

The problems to which we have thus far attended have been solved either by aid of the Differential Calculus alone, or the Integral Calculus alone. We now proceed to others that will require the aid of both.

III. RECTIFICATION OF CURVE LINES.

54. To rectify a curve line is to determine its length, or to assign a straight line which shall be equal to it, or to any proposed arc of it. When this can be done in a limited number of finite terms, the curve is said to be *rectifiable*.

PROB. *To determine the length of any arc of a curve in terms of the co-ordinates of its extremities.*

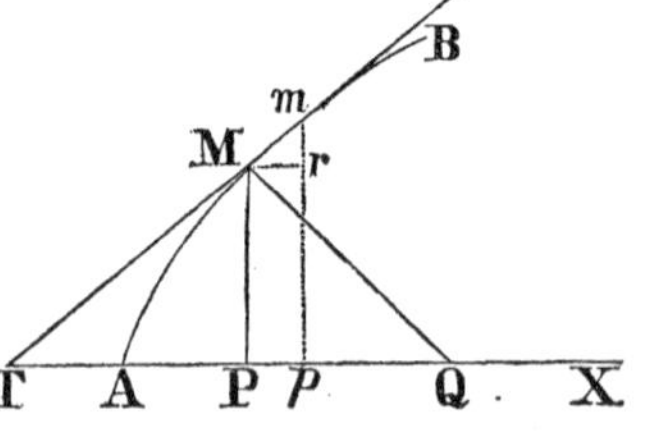

Let AM be the arc, $AP = x$ $PM = y$. Regarding the curve AM as a polygon of an infinite number of sides, let Mm be one of these sides; Mm will be the element or differential of the arc AM. Let $AM = z$, then $Mm = dz$.

Draw the ordinate pm, and draw Mr parallel to Ap; then M$r = dx$, $rm = dy$, and we have

$$dz = (dx^2 + dy^2)^{\frac{1}{2}},$$

for the differential of any arc of a curve.

55. As a first example let it be proposed to find an expression for the rectification of the parabola, whose equation is $y^3 = ax^2$.

Deducing the value of x from this equation, differentiating, and squaring, we obtain

$$dx^2 = \frac{9}{4} \cdot \frac{ydy^2}{a};$$

substituting in the general equation, we have

$$dz = \left(dy^2 + \frac{9}{4}\frac{ydy^2}{a}\right)^{\frac{1}{2}},$$

or

$$dz = dy\left(1 + \frac{9y}{4a}\right)^{\frac{1}{2}};$$

whence by integration

$$z = \frac{8a}{27}\left(1 + \frac{9y}{4a}\right)^{\frac{3}{2}} + C,$$

the *indefinite* Integral. To find the value of the constant, we reckon the arc from the vertex; which gives $y = 0$, $z = 0$; whence

$$\frac{8a}{27}(1)^{\frac{3}{2}} + C = 0,$$

or

$$C = -\frac{8a}{27};$$

and we have for the *corrected* integral

$$z = \frac{8a}{27}\left(1 + \frac{9y}{4a}\right)^{\frac{3}{2}} - \frac{8a}{27}.$$

If we now wish to determine the value of the arc between certain limits, $y = b$, $y = c$, for example, we put, in the indefinite integral, $y = b$, $y = c$, successively; and subtracting the former of these values from the latter we obtain the *definite* integral sought.

56. This curve is remarkable as being the first, the rectification of which was found. The discovery alone was sufficient to give celebrity to the name of Neil, an English Geometer, by whom, in about the year 1660, it was made. The problem is well adapted to illustrate the nature and power of the Calculus. As in the case of the problem of tangents, we remark,

1°. It is impossible to establish *directly* the relation between the arc, whose length is sought, and the co-ordinates of its extremities. Thus the resources of Algebra fail to furnish an equation for the problem.

2°. But the relation between the differential of the arc and the differentials of the co-ordinates, the auxiliaries introduced by the Calculus, is readily established; and thus an equation for the problem is very easily obtained. This is

$$dz = (dx^2 + dy^2)^{\frac{1}{2}}. \qquad (1)$$

3°. By aid of the *Differential Calculus* we are enabled to eliminate one of the indeterminates dx, which gives

$$dz = dy\left(1 + \frac{9y}{4a}\right)^{\frac{1}{2}}. \qquad (2)$$

4°. By aid of the *Integral Calculus*, we are then able to eliminate the remaining indeterminates dz, dy; and we thus obtain, finally, an equation free from the indeterminates, or auxiliaries employed, and in terms of the finite quantities given by the conditions of the problem. This is

$$z = \frac{8a}{27}\left(1 + \frac{9y}{4a}\right)^{\frac{3}{2}} - \frac{8a}{27}, \qquad (3)$$

and the problem is thus completely solved.

We see in this problem also the peculiar province of the Calculus. It consists, 1°. In the introduction of certain auxiliary quantities, in order to facilitate the forming the equations of the problem. 2°. In deriving the auxiliaries in such a manner from their primitives, the quantities under consideration in the problem, that, after having performed their office, their elimination, in virtue of the law of their derivation, is readily effected, and a solution of the problem is thus obtained in finite and determinate terms.

57. From what has been done, it will be perceived that there are three classes of questions which we have to consider in the Calculus. 1°. Those which can be solved by aid of the Differential Calculus alone. 2°. Those which can be solved by aid of the Integral Calculus alone. 3°. Those which require the aid of both; the Differential Calculus being necessary to prepare the equations for the application of the Integral Calculus. This latter class is by far the most numerous, or the general class.

IV. SURFACES OF SOLIDS.

58. We pass next to the quadrature of the surfaces of solids.

PROB. *Let the surface be that of a solid of revolution.*

Let AB be the generating curve, AX the axis of revolution. Regarding the curve as a polygon of an infinite number of infinitely small sides, let Mm, as before, be one of these sides. In the revolution of the arc Am, Mm will describe an infinitely small zone or surface which will be the element or differential of the surface of the solid.

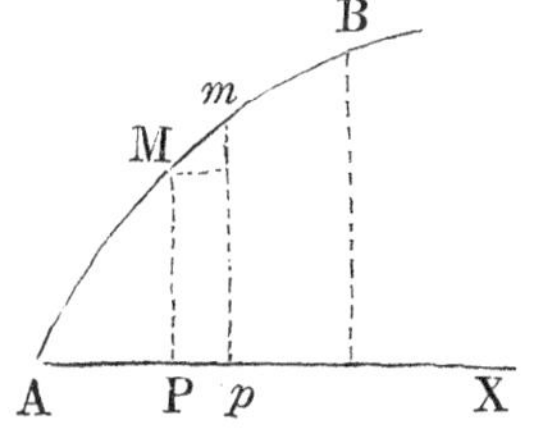

Employing the same notation as before, the circumference of the circle which has PM or y for its radius will be $2\pi y$. We have already found $Mm = (dx^2 + dy^2)^{\frac{1}{2}}$. We shall have, therefore, for the surface of the zone $2\pi y\,(dx^2 + dy^2)^{\frac{1}{2}}$. Putting S for the whole surface, we shall have, then, for the equation of the problem

$$dS = 2\pi y(dx^2 + dy^2)^{\frac{1}{2}}. \qquad (1)$$

59. Ex. 1. Let it now be proposed to find the surface of the pa-

raboloid, or the solid produced by the revolution of an arc AM, of a parabola about its axis.

From the equation of the parabola $y^2 = 2px$, we obtain

$$dx^2 = \frac{y^2 dy^2}{p^2}.$$

Substituting this in equation (1) we obtain

$$d\,S = \frac{2\pi}{p} y dy\,(y^2 + p^2)^{\frac{1}{2}}; \qquad (2)$$

whence by integration

$$S = \frac{2\pi}{3p}(y^2 + p^2)^{\frac{3}{2}} + C; \qquad (3)$$

and taking the integral between the limits $y = 0$, $y = b$, we have

$$S = \frac{2\pi}{3p}[(b^2 + p^2)^{\frac{3}{2}} - p^3]. \qquad (4)$$

Ex. 2. Let it be proposed next to find the surface of the sphere.

The equation of the circle, referred to the centre as the origin, is

$$x^2 + y^2 = R^2.$$

Differentiating, substituting and reducing, we obtain

$$d\,S = 2\pi R\,dx;$$

whence by integration $\quad S = 2\pi Rx + C.$

Taking the integral between the limits $x = 0$, $x = R$, we have

$$S = 2\pi R^2;$$

and between the limits $x = 0$, $x = -R$,

$$S = -2\pi R^2.$$

The sign — in this second expression, which is perfectly symmetrical with the first, shows merely that the two parts represented by them are situated on opposite sides of the plane passing through the centre. Adding these without reference to their signs, in order to obtain the whole surface, we have

$$2S = 4\pi R^2 = 2\pi R \times 2R,$$

that is, the surface of the sphere is equal *to four great circles*, or equal to the curved surface of the circumscribed cylinder.

Ex. 3. To find the surface of the parabolic conoid, the equation of the generating curve being $y^2 = 4ax$.

Ans. $\frac{8}{3}\pi a^{\frac{1}{2}}[(x+a)^{\frac{3}{2}} - a^{\frac{3}{2}}]$.

....

SECTION VI.

SUCCESSIVE DIFFERENTIALS, DIFFERENTIAL COEFFICIENTS.

60. We have now seen the aid which may be derived from the use of the auxiliaries already introduced by the Calculus. The same artifice, it is obvious, may be repeated at pleasure. Thus, when the nature of the problem is such, that an equation for it cannot readily be obtained by means of the auxiliaries first employed, we may regard these auxiliaries, in their turn, as primitives and introduce a new set of auxiliaries in relation to them, in the same manner that they have been introduced in relation to the primitive quantities which enter originally into the question; and so on. The only condition on which this successive and increasing number of auxiliaries may be employed is, that we have the means of their final elimination, so that the result shall be left in finite and determinate terms.

For a second set of auxiliaries, therefore, we employ the differentials of those first employed; the latter being regarded, in their turn, as variable, and composed also of elements infinitely small in relation to them. The differentials thus obtained are called *second differentials*. In like manner we employ, as the exigencies of the problem may require, the differentials of the second differentials, or the *third* differentials, and so on.

To indicate the second differential we repeat the letter d before its first differential, thus ddx; or which is better, by placing a figure

2 by the side of the d first employed, in the manner of an exponent. Thus d^2x indicates the second differential of x, and is read *second differential* of x. In like manner d^3x indicates the third differential of x; and so on.

61. Let us now see the manner in which these successive differentials are obtained.

The successive changes of a variable quantity may each be infinitely small and yet be unequal. For, as we have seen, though the magnitude of two infinitely small quantities cannot be assigned they may bear to each other any ratio whatever.

Let us, then, consider a variable quantity in three successive states infinitely near each other. If x represent the first, the three will be represented as follows

1st. state x
2d. $x + dx$
3d. $x + dx + dx + d(dx)$,

in which $d(dx)$ represents the differential of dx and is infinitely small in relation to it.

If we now subtract the first of these from the second, we have for the result dx. If we subtract the second from the third, the result will be $dx + d(dx)$. Thus we have for two successive differentials

$$dx, \text{ and } dx + d(dx);$$

subtracting next the first of these from the second in order to obtain the second differential, we obtain

$$d^2x = d(dx).$$

To obtain, therefore, the second differentials, *we differentiate the first differentials according to the rules already given.*

Ex. 1. Let it be proposed, as an example, to find the second differential of xy. Taking the first differential, we have

$$d(xy) = xdy + ydx;$$

regarding next dx and dy as variables as well as x and y, and differentiating again, we obtain

$$d^2(xy) = xd^2y + 2dxdy + yd^2x.$$

Ex. 2. To find the second differential of x^2. Proceeding in like manner we obtain

$$d^2(x^2) = 2xd^2x + 2dx^2.$$

The learner will carefully distinguish between the notation d^2x and dx^2; the former denoting that x has been differentiated twice, and the latter indicating the square of dx, or the square of the first differential.

62. We proceed in like manner to differentiate a quantity containing first differentials, whether itself the result of an exact differentiation or not.

Ex. 1. To find the differential of xdy.

$$d(xdy) = xd^2y + dxdy.$$

Ex. 2. Find the differential of $\frac{dy}{x}$. Ans. $\frac{d^2y}{x} - \frac{dxdy}{x^2}$.

Ex. 3. Find the differential of $\frac{dx}{dy}$. Ans. $\frac{d^2x}{dy} - \frac{dxd^2y}{dy^2}$.

We proceed in a manner altogether similar to find the third, fourth, &c. differentials.

63. The differential equation $dy = f'(x)dx$, derived from an equation $y = f(x)$ between two variables y and x, determines, as we have seen, only the relative value of the differentials dy and dx. We may impose, therefore, upon one or the other of these differentials any condition or restriction not incompatible with the equation by which they are connected. It is usual to assign a constant though indeterminate value to one of the differentials. This will greatly simplify the calculations in which these differentials occur. For the differential of a constant being 0, all the terms affected with the second differential of the variable, the first differential of which is regarded as constant, will be 0, or disappear.

Ex. Find the successive differentials of the function $y = ax^5$, on the supposition that dx has a constant value.

$$dy = 5ax^4dx$$
$$d^2y = 20ax^3dx^2$$
$$d^3y = 60ax^2dx^3$$
$$d^4y = 120axdx^4$$
$$d^5y = 120adx^5.$$

Here the operation terminates, since the last differential no longer contains x. In general the differentiation will terminate with that differential of the function which no longer contains the variable. Thus if x be the variable, ax^2 will admit of two differentiations, ax^3 of three, and so on.

In an equation between two variables, that is the *independent* variable whose first differential is assumed to be constant.

DIFFERENTIAL COEFFICIENTS.

64. In the expression $dy = 5ax^4dx$, found above, the coefficient $5ax^4$ determines the relation between the differential of the function and that of its variable; that is, dy is equal to $5ax^4$ times dx, or

$$\frac{dy}{dx} = 5ax^4.$$

In like manner the second differential of the function is $20ax^3$ times dx^2, or

$$\frac{d^2y}{dx^2} = 20ax^3.$$

The quantity which expresses the ratio between the differential of the function and that of its variable is called the *first differential coefficient*. That which expresses the ratio between the second differential of the function and the square of the differential of its variable is called the *second differential coefficient*; and so on. Thus $5ax^4$ is the first differential coefficient of $y = ax^5$, $20ax^3$ is the second differential coefficient, $60ax^2$ the third, and so on.

The successive differential coefficients are commonly read thus, *du* by *dx*, second *du* by dx^2; and so on.

Ex. 1. To find the successive differential coefficients of the function $u = ax^3 - bx^2 + cx - e$.

$$\text{Ans.}\quad \frac{du}{dx} = 3ax^2 - 2bx + c$$

$$\frac{d^2u}{dx^2} = 6ax - 2b$$

$$\frac{d^3u}{dx^3} = 6a.$$

Ex. 2. Find the fourth differential coefficient of $8x^4 - 3x^3 - 5x$.

Ans. 192.

Ex. 3. Find the third differential coefficient of the function $u = ax^n$. Ans. $n(n-1)(n-2)ax^{n-3}$.

In some cases, it is more convenient to employ the differential coefficients, instead of the differentials of the functions it is required to consider.

PARTIAL DIFFERENTIALS AND DIFFERENTIAL COEFFICIENTS.

65. We have sometimes occasion to differentiate a function with reference to a part only of the variables which enter into it. The result is called a *partial differential.* The differential coefficient, supposing one of the variables only to change its value, is called a *partial differential coefficient.*

Ex. 1. Let it be required to find the differential coefficients of the function $u = x^2y^2$, in relation to x and y severally.

Considering first x to vary and y to remain constant, and then y to vary and x to remain constant, we shall have for the partial differential coefficients sought

$$\frac{du}{dx} = 2y^2x \ (1), \qquad \frac{du}{dy} = 2x^2y. \quad (2)$$

2. Let it next be required to find the differential coefficients of these last in relation to y and x respectively.

Considering in the first x and dx as constant, and in the second y and dy as constant, we obtain

$$\frac{d^2u}{dxdy}=4yx \ (3), \quad \frac{d^2u}{dydx}=4xy. \quad (4)$$

The order in which the differentials appear in the denominator indicate the order in which the operations have been performed.

66. Since we return from the differential coefficient of a function to its differential by simply multiplying the coefficient by the differential of the variable, we may represent the differential of a function $u=f(x)$ by the notation $\frac{du}{dx}dx$. Thus the partial differentials of the function $u=f(x, y)$ will be expressed by $\frac{du}{dx}dx$, $\frac{du}{dy}dy$ respectively. And since the total differential of a function is equal to the sum of its partial differentials, we shall have for the total differential of a function $u=f(x, y)$

$$(du)=\frac{du}{dx}dx+\frac{du}{dy}dy, \qquad (1)$$

the du on the left being inclosed in a parenthesis to indicate that the total differential is meant.

67. Instead of the preceding notation, a partial differential may be indicated by placing, by the side of the d and a little below it, the letter with reference to which the differentiation is made. Thus d_xu indicates that u is differentiated with reference to x only; d_yu, that it is differentiated with reference to y only.

For the total differential we shall have d_xd_yu, &c. or more simply, (du). To indicate the differential coefficient we may write, in like manner, the letter c by the side of the d and a little below it. Thus d_cu represents, generally, the differential coefficient of u; $d_{c.x}u$ its differential coefficient with reference to x only.

A function u and its variable x are sometimes not given in the same equation. Thus let there be the two equations of the form

$u = f(y)$, $y = \varphi(x)$, for example. To find the differential coefficient of u in relation to x, the method which first presents itself is to eliminate the y between the equations, before applying the process of differentiation. The differential coefficient may be obtained, however, without this previous elimination in each particular case. Differentiating the equations $u = f(y)$, $y = \varphi(x)$, representing by $f'(y)$, $\varphi'(x)$ the coefficients of dy and dx respectively, substituting and dividing, we obtain

$$\frac{du}{dx} = f'(y) \times \varphi'(x)\text{; or, } \frac{du}{dx} = \frac{du}{dy} \cdot \frac{dy}{dx}. \quad (2)$$

The formula (2) will be found of frequent use hereafter. In general, it will be observed, that if three quantities are mutually dependent upon each other, *the differential coefficient of the first regarded as a function of the third, will be equal to the differential coefficient of the first regarded as a function of the second, multiplied by the differential coefficient of the second regarded as a function of the third.*

Ex. Let $v = bu^3$, $u = ax^2$. To find the differential coefficient of v considered as a function of x, we have

$$\frac{dv}{dx} = \frac{dv}{du} \cdot \frac{du}{dx} = 3bu^2 \times 2ax = 6abu^2x = 6a^3bx^5.$$

68. If in the equation $u = f(x, y)$, we have $u = 0$, then $du = 0$, and from equation (1) we have

$$\frac{du}{dx} dx + \frac{du}{dy} dy = 0\text{;}$$

hence

$$\frac{dy}{dx} = -\frac{du}{dx} \div \frac{du}{dy}. \quad (3)$$

By means of this formula we may readily find the differential coefficient of an implicit function, $f(x, y) = 0$, without resolving the equation. To apply the formula *we put the function* = u, *take the differential coefficients in relation to* x *and* y *respectively, divide the former by the latter and change the sign of the quotient.*

Ex. 1. Find the differential coefficient of the function

$$y^2 - 2mxy + x^2 - a^2 = 0.$$

Ans. $\frac{dy}{dx} = \frac{my - x}{y - mx}$.

Ex. 2. To find the differential coefficient of the function

$$y^3 - 3axy + x^3 = 0.$$

Ans. $\frac{ay - x^2}{y^2 - ax}$.

The differential coefficient, as in the preceding examples, will be usually expressed in terms of x and y. If the coefficient is desired in terms of x only, y may be eliminated by means of the original equation.

Ex. 3. To find the differential coefficient of the function $x^2 + y^2 - R^2 = 0$, in terms of x only.

Ans. $-\frac{x}{(x^2 - R^2)^{\frac{1}{2}}}$.

Ex. 4. To find the differential coefficient of the function

$$y^2 + 2xy + x^2 - a^2 = 0.$$

Ans. -1.

....

SECTION VII.

APPLICATION OF THE DIFFERENTIAL CALCULUS TO THE DEVELOPMENT OF FUNCTIONS.

69. A function is said to be developed, 1°. when the operations indicated are performed, or, 2°. when for these are substituted equivalent operations upon quantities more simple, and which follow each other according to some determinate law.

The means of developing by some simple process a complicated function, is a matter of great importance in the business of calculation. Algebra has furnished, in the Binomial Theorem, important aids to this purpose for all functions of the form $y = (a + x)^n$, n being any number entire or fractional, positive or negative.

By means of this theorem we obtain with great facility the development required in finite and determinate terms, when the exact development is possible; or, when this is not the case, in a series of such terms following each other by an obvious law, by means of which the development may be obtained with such degree of approximation as we please.

The Differential Calculus enables us to prepare other theorems for this same general object, more comprehensive and important. To some of these we shall now attend.

MACLAURIN'S THEOREM.

70. To begin with the most simple case, let $y = \phi(x)$, be a function of one variable. The function when developed must, it is evident, contain different powers of x combined with the constants which enter into it. As the most simple form, let us now see if the function can be developed according to the ascending powers of x.

Supposing the development possible, let us put

$$y = \mathrm{A} + \mathrm{B}x + \mathrm{C}x^2 + \mathrm{D}x^3 + \ldots\ldots \quad (1)$$

A, B, C, &c. being, respectively, the coefficients of x, which we are now to determine in such a manner that the development will be true, whatever value we assign to x.

Differentiating successively, and passing to the differential coefficients, we obtain

$$\frac{dy}{dx} = \mathrm{B} + 2\mathrm{C}x + 3\mathrm{D}x^2 + 4\mathrm{E}x^3 + \&c.$$

$$\frac{d^2y}{dx^2} = 2\mathrm{C} + 2 \,.\, 3\mathrm{D}x + 3 \,.\, 4\mathrm{E}x^2 + \&c.$$

$$\frac{d^3y}{dx^3} = 2 \,.\, 3\mathrm{D} + 2 \,.\, 3 \,.\, 4\mathrm{E}x + \&c.$$

Now since the development is required to be true for every value of x, it must be true for the particular value $x = 0$. Let then (y) represent what y becomes when $x = 0$, $\left(\frac{dy}{dx}\right)$ what $\frac{dy}{dx}$

becomes on the same hypothesis, and so on; we shall then have

$$(y)=\mathrm{A},\ \left(\frac{dy}{dx}\right)=\mathrm{B},\ \left(\frac{d^2y}{dx^2}\right)=2\mathrm{C},\ \left(\frac{d^3y}{dx^3}\right)=2\,.\,3\,.\,\mathrm{D},\ \&c.$$

whence by substitution in equation (1)

$$y=(y)+\left(\frac{dy}{dx}\right)x+\frac{1}{2}\left(\frac{d^2y}{dx^2}\right)x^2+\frac{1}{2.3}\left(\frac{d^3y}{dx^3}\right)x^3+\ldots.\ (2)$$

71. Such is the *Theorem of Maclaurin*. We will now apply it to a few particular cases.

Ex. 1. Let it be required to develop $y=\frac{1}{(a+x)}$ according to the ascending powers of x.

We obtain for the successive differential coefficients

$$\frac{dy}{dx}=-\frac{1}{(a+x)^2},\ \frac{d^2y}{dx^2}=\frac{2}{(a+x)^3},\ \frac{d^3y}{dx^3}=-\frac{2\,.\,3}{(a+x)^4}\ \&c.$$

whence making $x=0$, and substituting in the general formula (2), we obtain

$$\frac{1}{a+x}=\frac{1}{a}-\frac{x}{a^2}+\frac{x^2}{a^3}-\frac{x^3}{a^4}+\&c.$$

Ex. 2. Develop $y=(a^2+bx)^{\frac{1}{2}}$, according to the ascending powers of x.

We have for the successive differential coefficients

$$\frac{dy}{dx}=\frac{1}{2}(a^2+bx)^{-\frac{1}{2}}b$$

$$\frac{d^2y}{dx^2}=-\frac{1}{2}\cdot\frac{1}{2}(a^2+bx)^{-\frac{3}{2}}b^2$$

$$\frac{d^3y}{dx^3}=\frac{1}{2}\cdot\frac{1}{2}\cdot\frac{3}{2}(a^2+bx)^{-\frac{5}{2}}b^3$$

$$\&c. = \qquad \&c.$$

Putting in these results $x=0$, restoring the denominators, and substituting in the general formula (2), we obtain

$$(a^2+bx)^{\frac{1}{2}}=a+\frac{bx}{2a}-\frac{b^2x^2}{8a^3}+\frac{b^3x^3}{16a^5}-\&c.\ldots$$

Ex. 3. Let us take, as a third example, $y = (a + x)^m$.

We have for the successive differential coefficients

$$\frac{dy}{dx} = m(a + x)^{m-1},$$

$$\frac{d^2y}{dx^2} = m(m - 1)(a + x)^{m-2},$$

$$\frac{d^3y}{dx^3} = m(m - 1)(m - 2)(a + x)^{m-3}; \ \&c.$$

Substituting, as before, we obtain

$$(a + x)^m = a^m + ma^{m-1}x + m\frac{m-1}{2}a^{m-2}x^2 + \&c.$$

or the Binomial Theorem. Thus the theorem of Maclaurin comprehends the Binomial Theorem. We shall make more important applications hereafter. What has been done is sufficient to show the nature of the theorem and the utility of the Calculus in obtaining it.

The theorem of Maclaurin has been obtained on the supposition, that when $x = 0$, the function reduces to a *finite* quantity. When this is not the case the theorem cannot be employed for the development of the function.

This valuable theorem was discovered by Sterling in 1717, but appearing first in a work on Fluxions by Maclaurin, it has usually been attributed to him, and has gone by his name.

TAYLOR'S THEOREM.

72. As a second example of the use of the Calculus in the preparation of formulas for the development of functions, we proceed to investigate a theorem for the development of a function of the sum or difference of two variables.

In order to this we remark that, if we have a function of the sum or difference of two variables x and y, the differential coefficient will be the same, whether we suppose x to vary and y to remain constant, or y to vary and x to remain constant. For example let the function be

$$u = (x + y)^n.$$

Differentiating, supposing x to vary and y to remain constant, we have

$$\frac{du}{dx} = n(x + y)^{n-1}.$$

Differentiating in like manner, supposing y to vary and x to remain constant, we have

$$\frac{du}{dy} = n(x + y)^{n-1},$$

in which the differential coefficient is obviously the same, as on the first supposition.

This being premised, let $u' = f(x + y)$ be the proposed function, in which y is either positive or negative; and let $u = f(x)$ represent the value of the function when $y = 0$.

The development will consist, it is evident, of terms in x, y and constants; and it is proposed to obtain the development in terms of x and the ascending positive powers of y. Supposing the development possible, let us put

$$u' = \text{A} + \text{B}y + \text{C}y^2 + \text{D}y^3 + \text{\&c.} \qquad (1)$$

in which the coefficients A, B, C, &c. are independent of y, but dependent upon x and the constants which enter into the primitive function. It is now required to find for these coefficients values such, that the development will be true for all possible values which may be attributed to x and y.

Differentiating equation (1), on the supposition that x varies and y remains constant, and dividing by dx, we shall have

$$\frac{du'}{dx} = \frac{d\text{A}}{dx} + \frac{d\text{B}}{dx}y + \frac{d\text{C}}{dx}y^2 + \text{\&c.}$$

Differentiating again, supposing y to vary and x to remain constant, we obtain

$$\frac{du'}{dy} = \text{B} + 2\text{C}y + 3\text{D}y^2 + \text{\&c.}$$

But by the preceding principle we have

$$\frac{du'}{dx} = \frac{du'}{dy};$$

hence we shall have

$$\frac{dA}{dx} + \frac{dB}{dx}y + \frac{dC}{dx}y^2 + \&c. = B + 2Cy + 3Dy^2 + \&c. \quad (2)$$

But since the coefficients of this series are independent of y, and the equality exists whatever the value of y, it follows that the terms which involve the same powers of y in the two members must be respectively equal. We shall have, therefore,

$$\frac{dA}{dx} = B \ (3); \ \frac{dB}{dx} = 2C \ (4); \ \frac{dC}{dx} = 3D \ (5); \ \&c.$$

If now we make $y = 0$ in the proposed function $u' = f(x + y)$, since we have put $f(x) = u$, we shall have $A = u$.

Substituting for dA in (3) we have

$$B = \frac{du}{dx}; \qquad (6)$$

substituting next for dB in (4) its value derived from (6) we obtain

$$C = \frac{1}{1 \cdot 2} \frac{d^2u}{dx^2}.$$

In like manner we obtain

$$D = \frac{1}{1 \cdot 2 \cdot 3} \frac{d^3u}{dx^3};$$

and so on.

Substituting next in (1) the values of the coefficients A, B, C, &c. thus obtained, we shall have finally

$$u' = u + \frac{du}{dx}y + \frac{d^2u}{dx^2}\frac{y^2}{1 \cdot 2} + \frac{d^3u}{dx^3}\frac{y^3}{1 \cdot 2 \cdot 3} + \&c.$$

in which if y is negative the signs of the alternate terms will be changed. We have thus reached the theorem required.

This theorem was first given by Dr. Brook Taylor, and for that reason it is called *Taylor's Theorem*. It is of the greatest value in the development of functions. On another account, however, a higher interest is imparted to it. It is susceptible, though with far less facility, of being demonstrated by common Algebra. And being thus derived, Lagrange has made the whole structure of the Calculus to rest upon it.

73. We shall now show, in a few simple examples, the application of this theorem.

Ex. 1. To develop the function $u' = (x + y)^{\frac{1}{2}}$.

In this case u will be equal to $x^{\frac{1}{2}}$, and we shall have

$$\frac{du}{dx} = \tfrac{1}{2} x^{-\frac{1}{2}}, \quad \frac{d^2u}{dx^2} = -\tfrac{1}{4} x^{-\frac{3}{2}}, \quad \frac{d^3u}{dx^3} = \tfrac{3}{8} x^{-\frac{5}{2}}, \text{ \&c.}$$

Substituting these values in the theorem and restoring denominators, we obtain

$$(x + y)^{\frac{1}{2}} = x^{\frac{1}{2}} + \frac{1}{2}\frac{y}{x^{\frac{1}{2}}} - \frac{1}{8}\frac{y^2}{x^{\frac{3}{2}}} + \frac{1}{16}\frac{y^3}{x^{\frac{5}{2}}} -, \text{ \&c.}$$

Ex. 2. To develop the function $(x + y)^n$.

Ans. The same as by the Binomial Theorem.

Ex. 3. The theorem of Taylor may be applied to the development of the function $u = f(x)$, when x receives an arbitrary increment h and becomes $x + h$. To obtain the development we have only to substitute h for y in the formula.

74. In the demonstration of Taylor's Theorem we have supposed the coefficients A, B, C, &c. of y and its powers to be *finite* quantities, in order that the corresponding terms in both sides of equation (2) may be equated. If then such a value be assigned to the principal variable as to make any of these coefficients *infinite*, the theorem will not hold good; or, in other words, for that particular value it is said to *fail*.

Other theorems still more comprehensive, as those of Lagrange and Laplace, have been derived, by aid of the Calculus, for the development of functions. The subject is one of the greatest importance. But our limits will not permit us to proceed further.

SECTION VIII.

APPLICATION OF THE DIFFERENTIAL CALCULUS TO THE PROBLEM OF MAXIMA AND MINIMA.

75. We return to the problem of maxima and minima. In the examples thus far considered, we have assumed that the quantity proposed admits of a maximum or minimum, as the enunciation of the problem implies. The process, we have seen, is precisely the same, whether it be the one or the other which is required. Some means must, therefore, be found, by which to distinguish one of these cases from the other. To this we now proceed.

A quantity, it is obvious, may go on increasing continually, in which case it will have infinity for its maximum; or it may decrease continually, in which case it will have infinity negative for its minimum. It is usual, however, to exclude these cases and to define a maximum and minimum as follows.

1°. If a variable quantity gradually increases until it has reached a certain magnitude and then gradually decreases, at the end of the increase it is called a *maximum.*

2°. If a variable quantity gradually decreases until it has reached a certain magnitude and then gradually increases, at the end of the decrease it is called a *minimum.*

The characteristic of a maximum, then, is, that it is *greater* than the values which *immediately* precede and follow it; and of a minimum, that it is *less* than the values which *immediately* precede and follow it.

By means of this characteristic, when a quantity is susceptible of a maximum or minimum, we are able to determine whether it has the one or the other.

Ex. 1. Let the quantity be $8x - x^2$; to determine its maximum or minimum.

Applying the process already explained we have

$$d\,(8x - x^2) = 0\,;$$

whence $x = 4$. The proposed has, therefore, a maximum or minimum for $x = 4$. To determine which we substitute values for x, as follows

$$\begin{aligned}
\text{For } x &= 1, & 8x - x^2 &= 7 \\
x &= 2, & 8x - x^2 &= 12 \\
x &= 3, & 8x - x^2 &= 15 \\
x &= 4, & 8x - x^2 &= 16 \textit{ maximum.} \\
x &= 5, & 8x - x^2 &= 15 \\
x &= 6, & 8x - x^2 &= 12 \\
x &= 7, & 8x - x^2 &= 7.
\end{aligned}$$

Here, as we substitute successive values for x, the values of the proposed increase until we reach 16, after which they decrease. The value 16 is greater than those which immediately precede and follow it. The quantity is, therefore, a maximum when $x = 4$.

For a closer substitution let $x = 3.9$, for which we obtain $8x - x^2 = 15.99$; and again let $x = 4.1$, for which we obtain also 15.99, values very near 16, but still less as before.

For a still closer trial let $x = 3.99$, $x = 4.01$ respectively; the corresponding values of $8x - x^2$ will be each 15.9999. Thus, however close the substitution, 16 we find will still possess the characteristic property of a maximum.

Ex. 2. Let the quantity be $x^2 - 6x + 10$.

Applying the process and substituting values for x, as above, we find that the proposed has a minimum for $x = 3$.

MAXIMA AND MINIMA OF FUNCTIONS OF ONE VARIABLE.

76. We proceed next to maxima and minima of functions of one independent variable.

Every function of one independent variable, as $y = f(x)$, may be represented by a curve of which x and y are the corresponding co-ordinates.

Thus the function being represented by the curve (fig. 1), if the ordinate increases until it becomes PM and then decreases, PM will be a maximum. But if (fig. 2) the ordinate decreases until it becomes PM and then increases, PM will be a minimum.

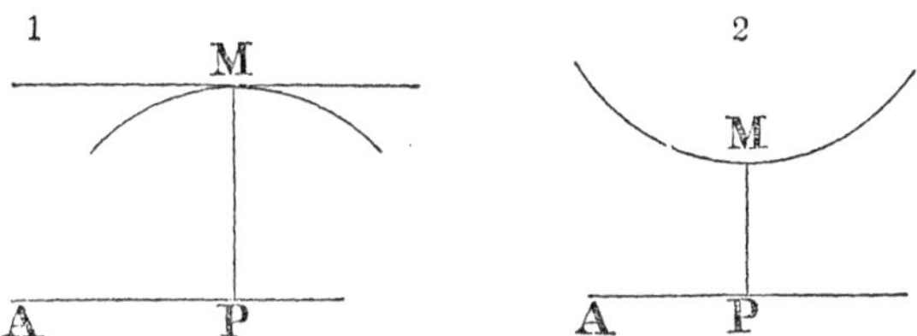

The same circumstances of increase and decrease may be repeated. Thus a function will have as many *maxima* or *minima* as the number of times it ceases to increase and begins to decrease, or ceases to decrease and begins to increase.

It will be observed, moreover, that when the ordinate PM becomes a maximum or minimum the tangent to the point M is parallel to the axis. Thus the question for finding the maxima and minima of functions is reduced to that of finding at what points on a curve the tangent is parallel to the axis. The general solution of the problem presents, therefore, all the difficulties of the problem of tangents of which it forms a part. The efforts for its solution in connection with that of tangents tended directly, especially in the labors of the great geometer Fermat, to the invention of the Calculus.

77. Resuming the solution of the problem under this point of view, the question is to find, on the curve that represents the function, the points at which the tangent is parallel to the axis.

We have seen, art. 34, that $\frac{dy}{dx}$, or the first differential co-efficient, expresses the value of the trigonometrical tangent of the angle which the tangent makes with the axis of x. In the case, then, of a maximum or minimum, we shall have

$$\frac{dy}{dx}=0\,;$$

since, when the tangent is parallel to the axis, the angle which it makes with this line is 0.

To determine, therefore, the maxima or minima values of a function $y = f(x)$ of a single variable, we put the first differential coefficient equal to 0, and find the roots of the equation thus formed. The roots of this equation will give all the values of x that will render y a maximum or minimum.

Ex. 1. Let us take, as an example, the function $y = x^2(5 - x)$, represented by the curve in the figure.

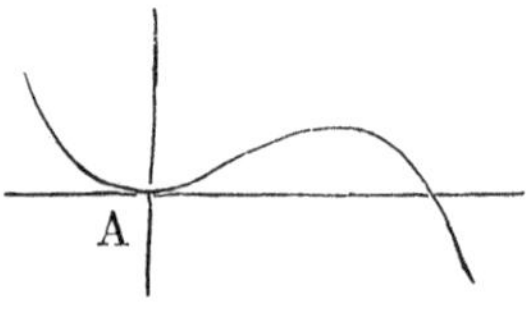

Differentiating this equation we obtain
$$\frac{dy}{dx} = x(10 - 3x).$$

Putting $x(10 - 3x) = 0$, we have for the roots of the equation $x = 0$, $x = 3\frac{1}{3}$.

Substituting $x = 0$ in the equation, we obtain $y = 0$. Substituting next $.1$ and $-.1$ for x, we obtain $.049$ for the value of y immediately following and $.051$ for that immediately preceding $x = 0$. The function, therefore, will be a *minimum* for $x = 0$. In like manner we find that it will be a *maximum* for $x = 3\frac{1}{3}$.

Ex. 2. Find the value of x which will render y a maximum in the equation $y = 10x - x^2$. Ans. $x = 5$.

Ex. 3. Find the value of x which will render y a maximum or minimum in the equation
$$y = x^3 - 18x^2 + 96x - 20.$$

Ans. $x = 4$ renders the function a maximum, and $x = 8$ renders it a minimum.

78. In the problem of maxima and minima there are, as we have seen, two questions to be considered.

1°. To determine when a proposed function is a maximum or minimum.

2°. To establish an easy criterion by which to distinguish the one from the other.

In the case of functions of one independent variable, the theorem of Taylor enables us to investigate a general method for these two objects, entirely independent of the theory of tangents.

Before proceding to this investigation, we remark, that in the development, by the theorem of Taylor, of the function $u = f(x)$, when x receives an arbitrary increment h; viz.

$$u' = u + \frac{du}{dx}h + \frac{d^2u}{dx^2}\frac{h^2}{1.2} + \frac{d^3u}{dx^3}\frac{h^3}{1.2.3} + \&c. \quad (1)$$

such a value may be assigned to the increment h that any one of the terms in the development shall be greater than the sum of all the terms which follow it.

Commencing with the second, the terms of the series may be written thus

$$\left(\frac{du}{dx} + \frac{d^2u}{dx^2}\cdot\frac{h}{1.2} + \frac{d^3u}{dx^3}\cdot\frac{h^2}{1.2.3} + \&c.\right)h.$$

Here, since, if h be taken equal to 0, all the terms within the parenthesis affected with h reduce to 0, h, it is evident, may be taken so small that the sum of all the terms affected with h shall be less than the term $\frac{du}{dx}$ which is independent of h. Putting this sum equal to M we shall have, then,

$$\frac{du}{dx} > M; \text{ whence } \frac{du}{dx}h > Mh;$$

that is, by taking h sufficiently small, the second term in the series (1) may become greater than the sum of all the terms which follow it. In like manner it may be shown that any term in the series may become greater than the sum of those which follow it.

Returning to our purpose, let the proposed function be $y = f(x)$, which we suppose now at its maximum or minimum. Let x be first diminished and then increased by a very small quantity h. Then $x - h$ will be the value which immediately precedes, and $x + h$ that which immediately follows x.

Designating the corresponding values of y by y', y'', we have for three contiguous values of the function

$$y' = f(x-h), \quad y = f(x), \quad y'' = f(x+h),$$

in which y', y'' are the same functions of $x-h$, $x+h$ that y is of x, and which may be calculated by Taylor's Theorem.

Developing by this theorem we obtain

$$y'' = y + \frac{dy}{dx}h + \frac{d^2y}{dx^2}\frac{h^2}{1\,.\,2} + \frac{d^3y}{dx^3}\frac{h^3}{1\,.\,2\,.\,3} + \&c.;$$

$$y' = y - \frac{dy}{dx}h + \frac{d^2y}{dx^2}\frac{h^2}{1\,.\,2} - \frac{d^3y}{dx^3}\frac{h^3}{1\,.\,2\,.\,3} + \&c.$$

Now in order that $y = f(x)$ may be a maximum or minimum, the values of y'' and y' must be both greater or both less than the value of y. That is, transposing y in the developments which gives

$$y'' - y = \frac{dy}{dx}h + \frac{d^2y}{dx^2}\frac{h^2}{1\,.\,2} + \frac{d^3y}{dx^3}\frac{h^3}{1.2.3} + \&c.,$$

$$y' - y = -\frac{dy}{dx}h + \frac{d^2y}{dx^2}\frac{h^2}{1\,.\,2} - \frac{d^3y}{dx^3}\frac{h^3}{1.2.3} + \&c.,$$

$y'' - y$ and $y' - y$ must be either both positive or both negative.

But h being taken such that any term in these two developments shall be greater than the sum of those which follow, the signs of the two, it is evident, will depend upon the sign of the first term in each or $\frac{dy}{dx}h$. And since the signs of this term are opposite in the two developments, the one being $+$ when the other is $-$, and the converse; therefore, in order that there may be a maximum or a minimum it is necessary that the value of x should be such as to render $\frac{dy}{dx}h$, or which is the same thing, $\frac{dy}{dx}$, the first differential coefficient, equal to 0. We shall have, then, for the condition sought

$$\frac{dy}{dx} = 0;$$

the roots of which will give all the values of x which, substituted in the function, will render the value of y a maximum or a minimum. Thus the first of the proposed questions is answered.

2. To determine next a criterion by which to distinguish the maxima from the minima.

The condition $\frac{dy}{dx} = 0$ being fulfilled, the developments above reduce to

$$y'' - y = \frac{d^2y}{dx^2}\frac{h^2}{1 \cdot 2} + \frac{d^3y}{dx^3}\frac{h^3}{1 \cdot 2 \cdot 3} + \&c.;$$

$$y' - y = \frac{d^2y}{dx^2}\frac{h^2}{1 \cdot 2} - \frac{d^3y}{dx^3}\frac{h^3}{1 \cdot 2 \cdot 3} + \&c.;$$

in which the signs of the sum of the terms will depend upon $\frac{d^2y}{dx^2}\frac{h^2}{1 \cdot 2}$, which has the same sign in both developments. If then the value for x which has reduced $\frac{dy}{dx}$ to 0 renders $\frac{d^2y}{dx^2}$ *negative*, the values of y'' and y' will both be less than y, and the function will be a *maximum*. But if the value for x renders $\frac{d^2y}{dx^2}$ *positive*, then the function will be a *minimum*.

To complete the investigation it is necessary to remark, that the value for x which reduces the first differential coefficient to 0, may reduce the second to 0 also. In this case, it is evident, in order that there may be a maximum or minimum, that the third differential coefficient must also reduce to 0. And this condition being fulfilled, we must then look to the fourth differential coefficient to determine whether the function is a maximum or minimum, in the same manner that we have already done in reference to the second, and so on.

The two objects proposed are now accomplished, and we have the following rule by which to determine the maxima and minima of a proposed function.

1°. *Find the first differential coefficient, and putting this equal to* 0, *find the roots of the equation thus formed.*

2°. *Substitute the roots successively in the second differential coefficient. Each root that gives a* negative *result will, when substituted in the function, make it a maximum; and each root which gives a* positive *result will make it a minimum.*

3°. *If either root reduces the second differential coefficient to* 0, *then substitute in the third, fourth, &c. until one is found which does not reduce to* 0. *If this differential coefficient be of an odd order, the root will not render the function either a maximum or minimum. But if it be of an even order and* negative *the function will be a maximum; if* positive *a minimum.*

79. We pass to some examples.

Ex. 1. To find the values of x which will render y a maximum or minimum in the function

$$y = x^3 - 9x^2 + 24x - 16.$$

Differentiating we have $\frac{dy}{dx} = 3x^2 - 18x + 24.$

Putting the second member equal to 0, and reducing,

$$x^2 - 6x + 8 = 0.$$

Resolving this last we obtain $x = 4$, $x = 2$.

For the second differential coefficient we have

$$\frac{d^2y}{dx^2} = 6x - 18.$$

Substituting 4 in this we obtain 6, a *positive* result. Substituting next 2 we obtain -6, a *negative* result. The proposed will have, therefore, a maximum for $x = 2$, and a minimum for $x = 4$.

Ex. 2. To find the values of x which will render y a maximum or minimum in the function

$$y = x^5 - 5x^4 + 5x^3 + 1.$$

Putting the first differential coefficient equal to 0, and dividing, we have

$$x^4 - 4x^3 + 3x^2 = 0,$$

or

$$x^2(x^2 - 4x + 3) = 0;$$

the roots of which are 0, 0, 1, 3.

Ans. There is a maximum for $x = 1$, and a minimum for $x = 3$.

Ex. 3. Find the maxima and minima values of the function

$$y = x^3 - 3x^2 - 24x + 85.$$

Ans. For $x = 4$, y is a minimum;
" $x = -2$, y is a maximum.

80. In the application of the above rule there are certain considerations which, in particular cases, serve to simplify the calculations.

1°. The same values of the variable which render the function a maximum or minimum, will render its product or quotient by a constant quantity a maximum or minimum. Hence, before applying the rule, *a constant factor may be struck out.*

2°. Any value of the variable which renders the function a maximum or minimum will render the second, third, and, in general, any power of it a maximum or minimum. Hence, *if a function is under a radical, the radical may be omitted.*

Ex. 1. Find the value for x which will render the function $y = 5(x - x^2)$ a maximum.

Suppressing the factor 5, we have

$$\frac{dy}{dx} = 1 - 2x;$$

from which we obtain $x = \frac{1}{2}$.

Ex. 2. Find the value for x which will render the function $y = (2ax - x^2)^{\frac{1}{2}}$ a maximum.

Omitting the radical sign, and putting the differential coefficient equal to 0, we obtain $2(a - x) = 0$;

whence $x = a$.

8

3°. It frequently happens that the value of the first differential coefficient may be decomposed into two factors, both of which contain x, and one only reducing to 0 for the particular value of x which renders the function a maximum or minimum. When this is the case, the process for finding the second differential coefficient, with a view to discover by its sign whether the function is a maximum or minimum, may be much simplified.

Let X and X′ be the two factors into which the first differential coefficient is decomposed, and both of which are functions of x; then

$$\frac{dy}{dx} = XX';$$

whence differentiating and dividing by dx,

$$\frac{d^2y}{dx^2} = \frac{XdX'}{dx} + \frac{X'dX}{dx};$$

let now X be the factor which we suppose equal to 0 for the value of x which renders the function a maximum or minimum; we shall, then, have

$$\frac{d^2y}{dx^2} = \frac{X'dX}{dx};$$

from which we obtain the following rule by which to find the second differential coefficient; *Find the differential coefficient of the factor in the first differential coefficient which reduces to* 0, *and multiply the result by the other factor.*

Thus let $\frac{dy}{dx} = \frac{x - a}{\sqrt{x}}$ be the first differential coefficient obtained from a function. To find the second differential coefficient for the value $x = a$, which renders the function a maximum or minimum, we write the proposed under the form

$$\frac{1}{\sqrt{x}}(x - a);$$

whence by the rule

$$\frac{d^2y}{dx^2} = \frac{1}{\sqrt{x}} \times \frac{d(x-a)}{dx} = \frac{1}{\sqrt{x}}.$$

Ex. 1. Find the values of x which will render the function $y = x^2(x-a)^6$ a maximum or minimum.

We have $$\frac{dy}{dx} = 2x(x-a)^6 + 6x^2(x-a)^5;$$
or reducing, putting equal to 0, and dividing by 2
$$x(x-a)^5(4x-a) = 0;$$
for which we have $x = 0,\ x = a,\ x = \frac{1}{4}a.$

For $x = \frac{1}{4}a$, the second differential coefficient, by the rule, is $4x(x-a)^5$, which is negative when we substitute $\frac{1}{4}a$ for x. Thus, there is a *maximum* for $x = \frac{1}{4}a$. In like manner we find a *minimum* for $x = a$, and $x = 0$.

Ex. 2. Find the values of x which will render y a maximum or minimum in the function
$$y = x^4 - 8x^3 + 22x^2 - 24x + 12 = 0.$$
Putting the first differential coefficient equal to 0, and dividing by 4, we find for the roots of the resulting equation, 1, 2, and 3.

Ans. For $x = 1$, y is a minimum;
$x = 2$, y is a maximum;
$x = 3$, y is a minimum.

81. The following examples will serve as a general exercise upon the preceding principles. Let the learner illustrate the results in the curves obtained by construction from the equations.

Ex. 1. Let $y = x(a-x)^2$.

Ans. There is a maximum for $x = \frac{1}{3}a$, and a minimum for $x = a$.

Ex. 2. Let $y = b + (x-a)^3$.

Ans. The function does not admit a maximum or minimum, since the third differential coefficient does not vanish with the second.

Ex. 3. Let $y = x^2(a-x)^2$.

Ans. There is a maximum for $x = \frac{1}{2}a$, and a minimum for $x = 0$, $x = a$.

Ex. 4. Let $y = a(x - b)^4$.

Ans. The fourth differential coefficient is $24a$, and there is a minimum for $x = b$.

Ex. 5. Let $y = x^3 - 18x^2 + 105x$.

Ans. There is a maximum when $x = 5$, and a minimum when $x = 7$.

82. We subjoin some additional problems to exhibit further the application of the Calculus to the important practical questions of the kind now under consideration.

1. To find the altitude of the greatest cylinder that can be inscribed in a given cone.

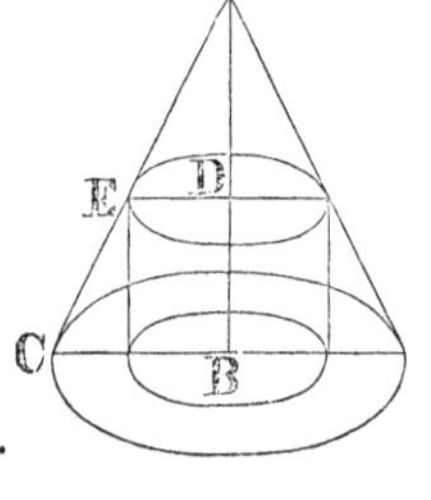

Suppose the cylinder inscribed, as in the figure, and let AB $= a$, BC $= b$, AD $= x$, ED $= y$; then BD $= a - x =$ the altitude of the cylinder, and we have for the solidity

$$V = \pi y^2 (a - x).$$

But from the similar triangles ADE, ABC, we obtain

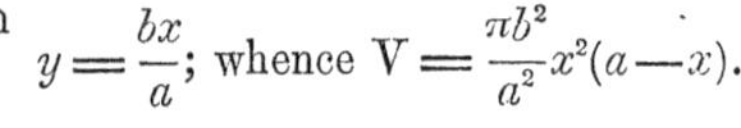

$$y = \frac{bx}{a}; \text{ whence } V = \frac{\pi b^2}{a^2} x^2 (a - x).$$

Ans. The altitude is one-third the altitude of the cone.

2. To find the greatest cylinder that can be inscribed in a given sphere.

Ans. The radius of the sphere being R, the radius of the cylinder $= R\sqrt{\frac{2}{3}}$.

3. What is the altitude of the maximum rectangle which can be inscribed in a given parabola?

Ans. Two-thirds the axis of the parabola.

4. Among all cones that can be inscribed in a given sphere, to determine that which has the greatest convex surface.

Let AMB be the generating semicircle, AM the slant side of the cone, PM the radius of the base. Let AB $= 2a$, AP $= x$, PM $= y$. Then PM $= (2ax - x^2)^{\frac{1}{2}}$, and AM $= (2ax)^{\frac{1}{2}}$; and, putting S for the convex surface, we have

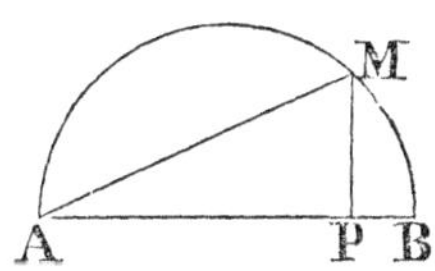

$$S = \pi (2ax - x^2)^{\frac{1}{2}} \times (2ax)^{\frac{1}{2}}.$$

Ans. The altitude of the cone is four-thirds the radius of the sphere.

5. What is the altitude of the maximum cone, which can be inscribed in a given paraboloid, the vertex of the cone being at the intersection of the axis with the base of the paraboloid?

Ans. Half the altitude of the paraboloid.

6. Of all cylinders of a given surface to determine that whose solidity is the greatest.

Ans. That in which the altitude is equal to twice the radius of the base.

7. To find the maximum cylinder inscribed in a given ellipsoid.

Ans. That in which the altitude is equal to the major axis divided by $\sqrt{3}$.

8. Of all right cones of a given convex surface to determine that whose solidity is greatest.

Ans. That in which the radius of the base is equal to the altitude divided by $\sqrt{2}$.

9. A privateer wishes to get to sea unobserved, but has to pass between two lights A and B, on opposite head lands. The intensity of the lights is known and the distance between them. At what point between them must she pass so as to be as little in the light as possible?

Let $a =$ the given distance AB; $b =$ the intensity of the light A at the unit of distance, $c =$ the intensity of B at the same distance. Let $x =$ the distance of the point sought from A. Then, since, by the principles of Optics, the intensities are inversely as the squares of the distances, we shall have for the illumination or intensity of light upon the point sought,

$$\frac{b}{x^2} + \frac{c}{(a-x)^2},$$

which by the question should be a minimum.

$$\text{Ans.} \quad x = \frac{ab^{\frac{1}{3}}}{b^{\frac{1}{3}} + c^{\frac{1}{3}}}.$$

10. The flame of a candle is directly over the centre of a circle, the radius of which is R; what ought to be its height above the plane of the circle, so as to illuminate the circumference as much as possible?

Let CB be the candle, B the position of the flame, and A any point in the circumference. By the principles of Optics, the intensity of the illumination of the plane at any point A, will be directly as the sine of the angle BAC, and inversely as the square of the distance AB. We shall have, therefore, for the effect of the candle to illuminate the point A,

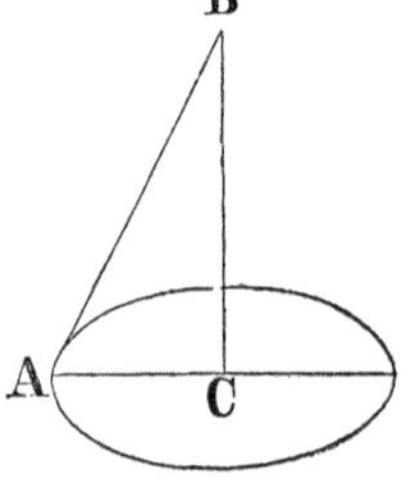

$$\frac{\sin \mathrm{BAC}}{\mathrm{AB}^2}.$$

Let $x = \sin \mathrm{BAC}$; then $\mathrm{AB} = \dfrac{\mathrm{R}}{\cos \mathrm{BAC}} = \dfrac{\mathrm{R}}{\sqrt{(1-x^2)}}$, whence

$$\frac{\sin \mathrm{BAC}}{\mathrm{AB}^2} = \frac{x(1-x^2)}{\mathrm{R}^2}.$$

We shall have, therefore, for the maximum, $x = \sqrt{\frac{1}{3}}$, or the sine of the angle $\mathrm{BAC} = 35° 16'$, nearly.

If we suppose $\mathrm{R} = 12$ inches, then the height of the candle will be 8.5 inches.

11. To find the point in the line joining the centres of two spheres, from which the greatest portion of spherical surface is visible.

The figure is a section through the centres C, C′ of the spheres. Let A be the point sought. Draw the tangents AM, Am, and the radii CM, C′m; and on CC′ let fall the perpendiculars MP, mp. Let CC′$=a$, r and $r'=$ the radii CM, C′m respectively; and let AC $=x$.

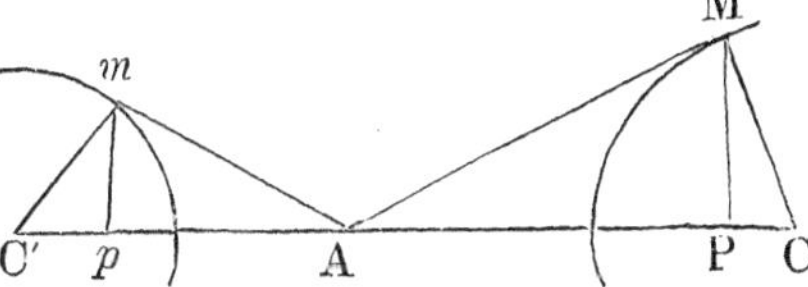

For the surface s of the sphere, which has r for its radius, visible from A, we shall have, art. 59,

$$s=2\pi r^2-2\pi r\times \text{PC}=2\pi r^2-2\pi\frac{r^3}{x};$$

and an equation for the problem will now be readily obtained.

$$\text{Ans.}\quad x=\frac{ar^{\frac{3}{2}}}{r^{\frac{3}{2}}+r'^{\frac{3}{2}}}.$$

SECTION IX.

APPLICATION OF THE CALCULUS TO THE THEORY OF CURVES.

83. We have seen in the Analytical Geometry the superior power of algebra compared with ordinary geometry, in all investigations relative to curve lines. For investigations of this description the Calculus is an instrument of far higher power. In the ordinary geometry the results obtained are usually derived from the consideration of particular curves and are limited to them. By the application of algebra to geometry we are enabled to classify the curves which come under consideration, and to obtain results applicable to all curves whatever.

The Calculus enables us to solve in most cases with far more facility, and with a higher generalization, the problems solved by the aid of algebra, while at the same time it enables us to obtain results which the ordinary algebra is utterly incompetent to reach.

In the use of this remarkable instrument, justly regarded as one of the most splendid conceptions of the human mind, geometry has, in fact, been brought to a state little short of absolute perfection.

We shall now, to the extent our limits permit, exhibit the power of the Calculus in its relation to investigations in which curve lines are concerned.

I. TANGENTS AND ASYMPTOTES.

84. The first problem in respect to curve lines is to determine a general method by which, when a curve is given by its equation, a tangent may be drawn to any point. We have already solved this problem for plane curves, referred to rectangular axes, by finding expressions for the subtangent and subnormal. By means of either of these expressions the tangent may be readily drawn. But the most elegant, as well as convenient solution of the problem is, to find an equation for the tangent line.

Let x, y be any point on a plane curve referred to rectangular axes; x' and y' the current co-ordinates of a right line referred to the same axes, and the same origin. In order that the right line may pass through the point x, y on the curve, we must have (An. Geom. art. 26)

$$y' - y = a(x' - x),$$

in which a is the trigonometrical tangent of the angle which the line makes with the axis of x. And in order next that the line may be tangent to the point x, y, we must have, as we have seen, art. 34,

$$a = \frac{dy}{dx};$$

whence by substitution

$$y' - y = \frac{dy}{dx}(x' - x), \qquad (1)$$

which is the differential equation of a tangent line to all plane curves expressed by rectangular co-ordinates.

Let the curve be the circle the equation to which, referred to the centre, is $x^2 + y^2 = \mathrm{R}^2$.

Differentiating the equation, deducing the value of $\frac{dy}{dx}$, substituting in equation (1), and reducing, we obtain

$$yy' + xx' = R^2, \qquad (2)$$

which is the equation, in finite terms, of a tangent line to the circle.

To apply this equation, let the radius of the circle be 5, and the co-ordinates of the point, through which the tangent is required to be drawn, be 4 and 3 respectively. By substitution in (2) we shall have

$$3y' + 4x' = 25, \qquad (3)$$

the equation of the tangent in this particular case, by means of which the tangent may be easily drawn.

The mode of applying the equation (1) is now obvious, as well as the extreme generality of the solution. We pass to a few additional examples.

Ex. 1. To draw a tangent line to any point on the curve $y = x - x^3$.

Pursuing the same course as above, we obtain for the equation of a tangent to the curve

$$y' - y = (1 - 3x^2)(x' - x).$$

As a particular case let $x = 0$, $y = 0$; then $y' = x'$. The tangent passes through the origin, and makes an angle of 45° with the axis of x.

Ex. 2. To find the equation of a tangent line to any point on the ellipse, the equation to which is $A^2y^2 + B^2x^2 = A^2B^2$.

$A^2yy' + B^2xx' = A^2B^2$. Ans.

Ex. 3. To find an equation for the tangent to the curve $xy = a^2$.

Ex. 4. To find an equation for the tangent to the curve $y^2 = 4ax$.

85. To construct a tangent line it is in general, as we have seen, most convenient to find the points in which the line intersects the axes.

Putting, successively, in the general equation (1), $y'=0$, $x'=0$, we shall have for the points of intersection

$$x'=x-y\frac{dx}{dy};\quad y'=y-x\frac{dy}{dx}.$$

From the former of these expressions we have

$$x-x'=y\frac{dx}{dy}=\text{PT};\ \text{(fig. art. 30)}$$

the subtangent to any point x, y on the curve ; the same as obtained in art. 30.

86. Since the normal makes a right angle with the tangent, we must have (An. Geom. art. 28) the relation $aa'+1=0$, a and a' representing the trigonometrical tangents of the angles which these lines make with the axis of x respectively. Hence we have

$$y'-y=-\frac{dx}{dy}(x'-x),\qquad (1)$$

for the differential equation of the normal to any point on the curve.

If in this equation we put $y'=0$, we shall have

$$x'-x=y\frac{dy}{dx}=\text{PQ},$$

the subnormal for any point on the curve ; the same as in art. 33.

87. We shall now examine some particular cases in the general problem of tangents.

1. To determine the points on a curve at which the tangent is parallel to the axis of x.

Since, when the tangent is parallel to the axis of x, the angle it makes with this line is 0, we shall have, as we have already seen,

$$\frac{dy}{dx}=0.$$

Ex. 1. To find on the circle the point where the tangent is parallel to the axis.

From the equation of the circle $y^2 = 2ax - x^2$, we obtain

$$\frac{dy}{dx} = \frac{a - x}{y};$$

and the tangent, it is evident, will be parallel to the axis of x for the value $x = a$, or at the extremity of the ordinate to the centre.

Ex. 2. To find on the curve $y = x^3 - 3x^2 - 24x + 85$, the points at which the tangent is parallel to the axis of x.

Ans. At the points for which $x = 4$, $x = -2$.

2. To determine the points on a curve, at which the tangent is perpendicular to the axis of x.

Since, when the tangent is perpendicular, the angle is equal to $90°$, we shall have

$$\frac{dy}{dx} = \infty.$$

Ex. 1. To find on the circle the point where the tangent is perpendicular to the axis of x.

Ans. For the point $y = 0$, or the origin.

Ex. 2. To find on the curve $y = a + 2(x - b)^{\frac{2}{3}}$, the point at which the tangent is perpendicular to the axis of x.

Ans. At the point for which $x = b$.

3. To find next the points on the curve at which the tangent makes any given angle with the axis of x.

Ex. 1. Let it be required to find on the common parabola $y^2 = 2px$ the point where the tangent makes an angle of $45°$ with the axis of x.

From what has been said we have, since tangent $45° = 1$,

$$\frac{dy}{dx} = 1.$$

Substituting for this value in the equation, we obtain $y = p$; whence $x = \frac{1}{2}p$. When the angle, therefore, is $45°$, the ordinate to the point of tangency will pass through the focus.

Ex. 2. The circle being referred to the centre, to find the point where the tangent makes an angle of 45° with the axis of x.

Ans. At $x = \dfrac{R}{\sqrt{2}}$.

4. To find an expression for the length of the tangent, we have, (fig. art. 30)

$$TM = (TP^2 + PM^2)^{\frac{1}{2}};$$

whence

$$TM = y\left(1 + \frac{dx^2}{dy^2}\right)^{\frac{1}{2}}.$$

In like manner we obtain for the length of the normal

$$QM = y\left(1 + \frac{dy^2}{dx^2}\right)^{\frac{1}{2}}.$$

By means of these formulas the length of the tangent and normal to any point may be easily calculated.

5. To find the angle which a curve makes with the axis of x, since the curve and tangent coincide at the point of tangency, it follows that the inclination of a curve to its axis at any point will be the same as that of the tangent at that point.

Asymptotes.

38. In what precedes we have supposed the point, through which the tangent is required to be drawn, at a finite distance from the origin. But in the case of a curve which has infinite branches, the point of tangency may be regarded as at an infinite distance, and we may wish to determine the tangent for this particular case.

The tangent to a point on a curve at an infinite distance from the origin, is called an *asymptote*. An asymptote, therefore, may be regarded as a line which continually approaches a curve, but which it can never reach except at an infinite distance.

A plane curve being proposed we have then, 1°. To ascertain whether the curve has a rectilinear asymptote. 2°. To determine its position.

Let AX, AY be the co-ordinate axes, and let us resume the expressions for the distances of the points T and D in which the tangent MT intersects the axes, viz.

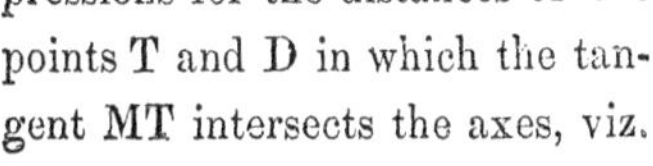

$$x' = x - y\frac{dx}{dy} = \text{AT};$$

$$y' = y - x\frac{dy}{dx} = \text{AD}.$$

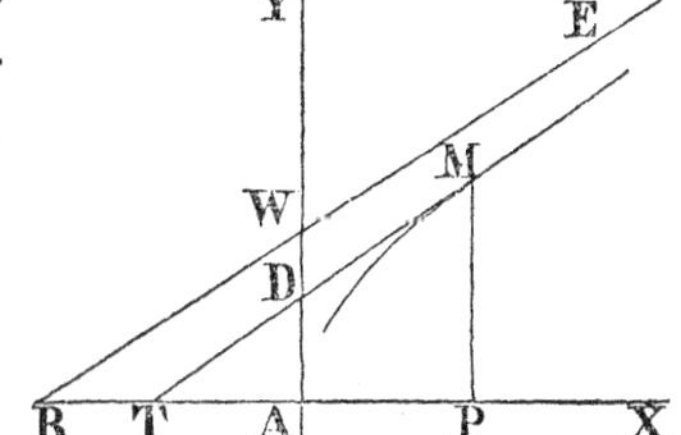

Let RE be the asymptote; as the point M moves along the curve the tangent MT will approach more nearly the asymptote RE, with which it will coincide when M is taken at an infinite distance. In this case AD will become equal to AW, and AT to AR. Whence to find whether the curve has an asymptote, we determine, from its equation, the values of AT and AD. If then the hypothesis x, or $y = \infty$ renders either the value of AT or AD finite, the curve, it is evident, will have an asymptote.

To determine the position of the asymptote we find the values of AT, and AD on the hypothesis x, or $y = \infty$, and we shall thus have the points T, and D in which the asymptote intersects the axes.

Several cases, it is obvious, will occur in the position of the asymptote.

1°. If the values of AT and AD are both finite, the asymptote will intersect both the co-ordinate axes.

2°. If one of the values is finite and the other infinite, the asymptote will be parallel to one of the co-ordinate axes.

3°. If both values become 0, the asymptote will pass through the origin of co-ordinates.

In this last case, but one point of the asymptote will be determined, but its direction may be found by determining the value of $\frac{dy}{dx}$, on the supposition that the co-ordinates are infinite.

89. We take next some examples.

Ex. 1. To determine whether the curve $y^2 = 2x + 3x^2$ has asymptotes.

Resuming the equations for AT, AD, we have

$$\text{AT} = x - y\frac{dx}{dy}, \quad \text{AD} = y - x\frac{dy}{dx},$$

whence, differentiating the equation, substituting and reducing, we obtain

$$\text{AT} = -\frac{x}{1+3x}, \quad \text{AD} = \frac{x}{(2x+3x^2)^{\frac{1}{2}}};$$

which become, supposing $x = \infty$,

$$\text{AT} = -\frac{1}{3}, \quad \text{AD} = \frac{1}{\pm\sqrt{3}}.$$

Thus the curve has two asymptotes which intersect the axes at

$$x = -\frac{1}{3}, \ y = \frac{1}{\pm\sqrt{3}}.$$

Ex. 2. Let the equation to the curve be $y^3 = ax^2 + x^3$.

The curve has an asymptote which intersects the axis of y at a distance $\frac{1}{3}a$, and that of x at a distance of $-\frac{1}{3}a$ from the origin.

Ex. 3. Let the equation to the curve be $y^2 = 4x^2 - 4$.

The curve has two asymptotes which pass through the origin, and make with the axis of x an angle for which 2 is the trigonometrical tangent.

II. DIRECTION OF CURVATURE.

90. We proceed next to inquire how the position of a curve in respect to the axis of abscissas may be found; or, in other words, to determine for any point whether the curve is convex or concave toward this axis.

M being any point of the curve, let $\text{AP} = x$, $\text{PM} = y$. Suppose x to receive an arbitrary increment PP′, and draw the ordinate P′M′. Through M draw the tangent MT meeting P′M′ in T, and draw also MR parallel to AX. Putting $\text{PP}' = h$, the ordinate P′M′ will, it is evident, be the same function of $x + h$,

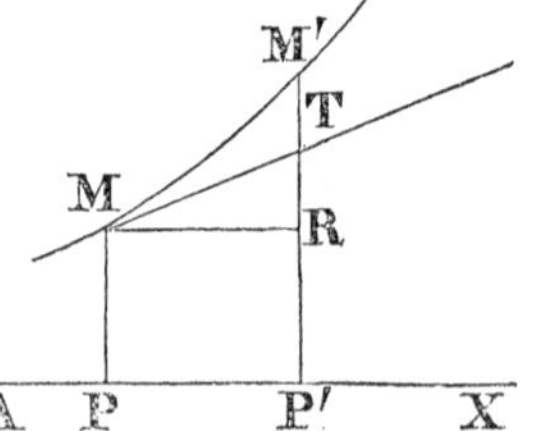

that PM is of x; whence by Taylor's Theorem

$$P'M' = y + \frac{dy}{dx}h + \frac{d^2y}{dx^2}\frac{h^2}{1 \cdot 2} + \&c. \qquad (1)$$

wherefore, $RM' = P'M' - PM = \frac{dy}{dx}h + \frac{d^2y}{dx^2}\frac{h^2}{1 \cdot 2} + \&c.$

But $RT = MR \times \text{Tan. } TMR = \frac{dy}{dx}h$;

whence $RM' = RT + \frac{d^2y}{dx^2}\frac{h^2}{1 \cdot 2} + \&c.$

and we have $TM' = RM' - RT = \frac{d^2y}{dx^2}\frac{h^2}{1 \cdot 2} + \&c.$

If the curve lies above its tangent, or is convex toward the axis of x, TM′, it is evident, will be positive. But h being altogether arbitrary, may be taken so small that the first term in the value of TM′ will be greater than the sum of all the succeeding terms. Thus the sign of TM′ will depend upon that of $\frac{d^2y}{dx^2}\frac{h^2}{1.2}$. Hence, if the curve is convex toward the axis, this term must be positive; or, since h^2 is necessarily positive, $\frac{d^2y}{dx^2}$ or the second differential coefficient, must be positive.

The contrary will be the case if the curve is concave toward the axis; for the curve then falling below the tangent, TM′ will be negative. Thus a curve will be convex or concave toward the axis of x, according as the second differential coefficient of the ordinate is *positive* or *negative*.

In what has been done we have supposed the ordinate y to be positive. If it is negative, the curve being situated below the axis of x, the contrary results will be obtained. We conclude, therefore, more generally, *That a curve will be convex or concave toward the axis of abscissas, according as the ordinate and its second differential coefficient have the same or contrary signs.*

Ex. 1. To determine whether the circumference of a circle is convex or concave toward the axis of abscissas.

Resolving and differentiating the equation $x^2 + y^2 = R^2$, we obtain

$$\frac{d^2y}{dx^2} = -\frac{R^2}{y^3},$$

which is negative when y is positive, and positive when y is negative. The circumference is, therefore, concave toward the axis of abscissas.

Ex. 2. Given the curve $y = b + c(x - a)^2$, to find the direction of the curvature. Ans. Convex toward the axis of x.

Ex. 3. Given the curve $y = a^2 \sqrt{(x - a)}$, to find the direction of the curvature. Ans. Concave toward the axis of x.

Ex. 4. Given the curve $y = a^2(a - x)^3$ to find the direction of the curvature.

III. SINGULAR POINTS OF CURVES.

91. In the examination of a curve we are led to notice certain points which possess some peculiar property not possessed by the points immediately adjacent, and which, on this account, are called *Singular Points.*

We have, in the course of the preceding investigations, incidentally determined several points of this description. They are

1°. *Points where the tangent is parallel to, or perpendicular to the axis.*

2°. *Points of maximum or minimum ordinates.*

We proceed to determine the characteristics of such other singular points as are ordinarily to be found in plane curves.

Points of Inflexion.

92. When a curve changes the direction of its curvature, as from being concave toward the axis to becoming convex, or the converse, the point at which the change takes place is called a point of *inflexion or contrary flexure.*

If a curve is convex toward the axis of x, the ordinate and second differential coefficient have each the same sign, if concave, they have contrary signs. When a curve, therefore, changes the direction of its curvature, the second differential coefficient changes sign. But this coefficient can change its sign only by passing through 0 or infinity. For a point of inflexion we must have, therefore,

$$\frac{d^2y}{dx^2}=0, \text{ or } \frac{d^2y}{dx^2}=\infty,$$

the roots of which will give all the values of x for which there is a point of inflexion.

Let $x=a$ be one of these roots. If on substituting successively in the coefficient a value for x a little greater and a little less than a, the coefficient changes sign, there will be a point of inflexion for $x=a$.

Ex. 1. To determine whether the curve $y=b+(x-a)^3$ has a point of inflexion.

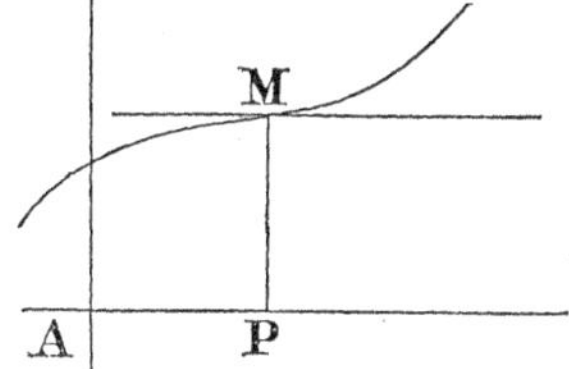

We find for the second differential coefficient

$$\frac{d^2y}{dx^2}=6(x-a).$$

Here the value $x=a$ renders the coefficient 0; and if we substitute for x a value a little greater than a, the coefficient will be positive, if a little less, the coefficient will be negative. Thus the coefficient changes sign at $x=a$; and there is a point of inflexion for $x=a$.

The tangent to the point $x=a$ is parallel to the axis of x, since for $x=a$ the first differential coefficient also reduces to 0. On the left of the point of inflexion the curve is concave toward the axis of x; on the right it is convex.

Ex. 2. To determine whether the curve $y=(a-x)^{\frac{5}{3}}+ax$ has a point of contrary flexure.

We have for the second differential coefficient

$$\frac{d^2y}{dx^2}=\frac{10}{9(a-x)^{\frac{1}{3}}}.$$

Here if $x = a$, the second differential coefficient will be infinite. For x less than a, it will be positive; for x greater than a, it will be negative. There is, therefore, a point of inflexion for $x = a$.

Ex. 3. The curve $a^2y = x^3 - cx^2$, has a point of inflexion at $x = \frac{1}{3}c$.

Ex. 4. The curve $y = x + 36x^2 - 2x^3 - x^4$, has a point of inflexion at $x = 2$, and $x = -3$.

Multiple Points.

93. A point in which two or more branches of a curve touch or intersect each other is called a *multiple* point.

If two branches meet at the same point, it is called a *double* point; if three, a *triple* point, and so on.

As each branch of the curve has its tangent, there will be at a multiple point as many tangents as there are branches of the curve which meet in the point. At a multiple point, therefore, the ordinate will have but one value, while the first differential coefficient will have a number of values, according to the multiplicity of the point.

Ex. 1. To determine whether the curve $y = (x - a)\sqrt{x} + b$, has a multiple point.

For the first differential coefficient we find

$$\frac{dy}{dx} = \pm \frac{3x - a}{2\sqrt{x}}.$$

The curve has a double point, since there are two values for $\frac{dy}{dx}$.

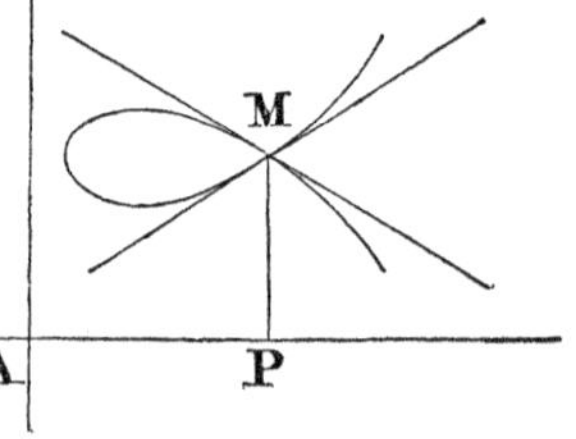

To determine the position of the double point, it will be observed that y will have two values for every value of x greater or less than a. For $x = a$ we have $y = b$, which gives the position of the double point.

To determine the tangents, we substitute in the first differential coefficient $x = a$, which gives

$$\frac{dy}{dx} = \pm \sqrt{a}.$$

The tangents, therefore, make with the axis of x angles, the trigonometrical tangents of which are $x = + \sqrt{a}$, $x = - \sqrt{a}$, respectively.

Ex. 2. Given the curve $y^2 = x^2 - x^4$, to determine whether it has a multiple point.

Ans. There is a multiple point at the origin; at this point the tangents make angles with the axis of x, the trigonometrical tangents of which are ± 1 respectively.

Ex. 3. $y^2 = (x-1)^2 x$. There is a multiple point at $x = 1$.

If the equation of a curve be freed from radical quantities, the differential coefficient corresponding to a multiple point will generally reduce to $\frac{0}{0}$, the ordinary sign of indetermination. And in this case to determine the tangents we must find the true values of this expression.

Cusps.

94. A multiple point at which two or more branches of a curve stop, and at which they have a common tangent, is called a *cusp*.

There are two species of cusps, 1°, when the two branches at the point of contact lie (fig. 1) upon opposite sides of the tangent; this is called a cusp of the *first kind*. 2°, when the two branches lie (fig. 2) upon the same side of the tangent; this is called a cusp of the *second kind*.

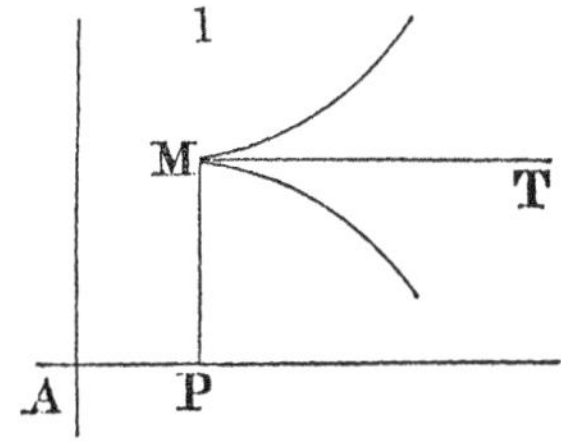

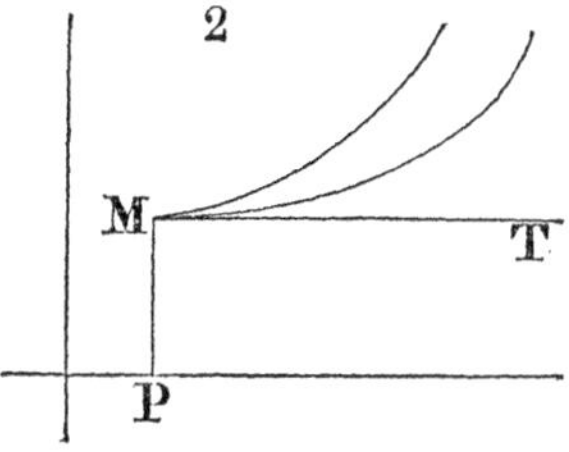

The conditions necessary to a cusp are

1°. For a particular value, a, of x, the ordinate y and the first differential coefficient must have but one value.

2°. Substituting successively for x, $a+h$, and $a-h$, for one of these values, h being taken sufficiently small, y is impossible, while for the other the second differential coefficient will have two different values.

Ex. 1. To determine whether the curve $(y-b)^2=(x-a)^3$ has a cusp.

Resolving the equation we obtain $y=b\pm(x-a)^{\frac{3}{2}}$;
and by differentiation

$$\frac{dy}{dx}=\frac{3}{2}(x-a)^{\frac{1}{2}}, \qquad \frac{d^2y}{dx^2}=\frac{3}{4(x-a)^{\frac{1}{2}}}.$$

The curve is represented in figure 1 above.

1°. For $x=\mathrm{AP}=a$, $y=\mathrm{PM}=b$. There is but one value of y, therefore, for $x=a$. Again for $x=a$, $\frac{dy}{dx}=0$. There is, therefore, but one tangent MT to the point M, which is parallel to the axis of x. 2°. For values of x less than a, the values of y are imaginary. Thus the curve stops at M. 3°. For values of x greater than a the second differential coefficient has two values, one $+$, the other $-$. The curve has, therefore, two branches which extend indefinitely to the right of M, and the cusp is of the first kind.

Ex. 2. $y=a+x+bx^2+cx^{\frac{5}{2}}$. The curve has a cusp at $x=0$, $y=a$. The tangent is inclined to the axis of x at an angle of 45°.

Ex. 3. $y^2=x^2(1-x^2)^3$. The curve has two cusps of the first kind, corresponding to $x=\pm 1$.

Ex. 4. $y=a+2(x-b)^{\frac{2}{3}}$. For this curve, as it will be easy to discover, there is a cusp for $x=b$, $y=a$, although, on applying the process above, the first condition only is fulfilled. If, however, we resolve the equation in relation to x, we find

$$x=b+\frac{(y-a)^{\frac{3}{2}}}{\sqrt{8}};\ \frac{dx}{dy}=\frac{3(y-a)^{\frac{1}{2}}}{2\sqrt{8}};\ \frac{d^2y}{dx^2}=\frac{3}{4\sqrt{8}(y-a)^{\frac{1}{2}}};$$

by which we find a cusp at $y=a$, $x=b$, as there should be.

To determine all the cusps which belong to a curve, the process above employed with reference to the axis of x, must, it is obvious, be repeated with reference to the axis of y.

Isolated or Conjugate Points.

95. The equation of a curve will sometimes give a point that is detached from all the rest. This is called an *isolated or conjugate point.*

Thus in the equation $y=(a+x)\sqrt{x}$, if x is negative, y will, in general, be imaginary. But for the value $x=-a$, y will be real and equal to 0. For positive values of x there will be two values of y. Thus the curve will consist of two branches extending indefinitely to the right of the origin; while on the left for $x=-a$, $y=0$, there will be a solitary point unconnected with any other.

IV. TRACING OF CURVES.

96. From what has been done we see the course to be pursued in tracing out a curve expressed by an equation. It is in general as follows.

1°. Resolve the equation in relation to one of the variables, y for example.

2°. Assign such positive values to x as will be sufficient to determine those for which $y=0$, $y=\infty$, or y impossible. The first will give the points in which the curve intersects the axis of x, the second determines the infinite branches, and the third shows where the curve stops in its course. Assign also for the same object negative values to x, and deduce the values of y corresponding. In both cases attend to the positive and negative values of y, so as to obtain the branches on both sides of the axis of abscissas.

3°. Find whether the curve has asymptotes, and determine their position if they exist.

4°. Find the first differential coefficient, and determine by means of it the maximum and minimum ordinates, the angle at which the curve cuts the axis, and its multiple points if any exist.

5°. Find the second differential coefficient, and thence determine the direction of the curvature of the different branches, and the points of inflexion if there be any.

6°. Find the remaining singular points if any exist.

Ex. 1. To trace the curve represented by the equation

$$y = \frac{x}{1 + x^2}.$$

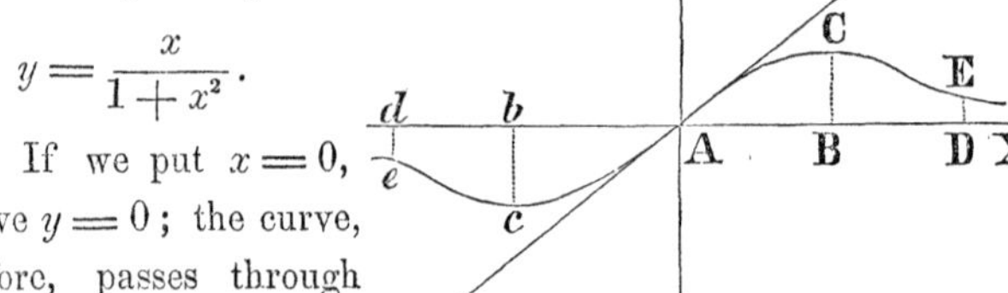

1°. If we put $x = 0$, we have $y = 0$; the curve, therefore, passes through the origin.

2°. Let x be positive, then y will be positive for every value of x from 0 to ∞. If x be *negative* then y will be negative for every value of x from 0 to $-\infty$. The curve has, therefore, two branches situated on opposite sides of the axis of x.

3°. The value $x = \infty$, or $x = -\infty$, reduces the value of y to 0; the axis of x will be, therefore, an asymptote to the curve in the direction both of x positive and x negative.

4°. For the first differential coefficient we have

$$\frac{dy}{dx} = \frac{1 - x^2}{(1 + x^2)^2}.$$

Putting this equal to 0, we obtain $x = \pm 1$. Thus we have two maximum ordinates, viz. at $x = +1$, $x = -1$, for which $y = \frac{1}{2}$, $y = -\frac{1}{2}$ respectively.

If we put $x = 0$, then $\frac{dy}{dx} = 1$, and the curve, or its tangent, intersects the axis at an angle of 45°.

5°. For the second differential coefficient we obtain

$$\frac{d^2y}{dx^2} = -\frac{2x(3-x^2)}{(1+x^2)^3}.$$

Putting this equal to 0, we have $x(3-x^2)=0$, which gives $x=0$, $x=\pm\sqrt{3}$. The second differential coefficient and y will have, therefore, the same or opposite signs according as x is greater or less than $\sqrt{3}$. Thus there will be points of inflexion at $x=+\sqrt{3}$, $x=-\sqrt{3}$, for which we have $y=\frac{1}{4}\sqrt{3}$, $y=-\frac{1}{4}\sqrt{3}$.

From $x=-\sqrt{3}$ to $x=+\sqrt{3}$ the curve is concave to the axis of x. There will be, therefore, also a point of inflexion at the origin. From $x=\sqrt{3}$ to $x=\infty$, and also from $x=-\sqrt{3}$ to $x=-\infty$, the curve is convex toward the axis.

To trace the curve we take AX, AY for the axes and from A we set off $x=\text{AB}=1$, $x=\text{A}b=-1$, and erecting the perpendiculars $\text{BC}=\frac{1}{2}$, $bc=-\frac{1}{2}$, we have the points of the maximum ordinates through which the curve must pass. In like manner we construct the points E, e of contrary flexure. Then, since the curve passes through the origin, and the direction of its curvature is known between all the singular points, it may be easily traced as in the figure.

Ex. 2. To trace the curve $y=x^3-6x^2+11x-6$.

1°. The curve cuts the axis of y at $y=-6$, and the axis of x, at $x=1$, $x=2$, $x=3$.

2°. It has maximum ordinates at $x=2+\sqrt{\frac{1}{3}}$, $x=2-\sqrt{\frac{1}{3}}$.

3°. Between $x=0$ and $x=1$ it is convex toward the axis of x; between $x=1$ and $x=3$ it is concave, beyond which it is convex. There is a point of inflexion at $x=2$.

Ex. 3. To trace the curve $(y-x^2)^2=x^5$.

Resolving the equation we have

$$y=x^2\pm x^{\frac{5}{2}}.$$

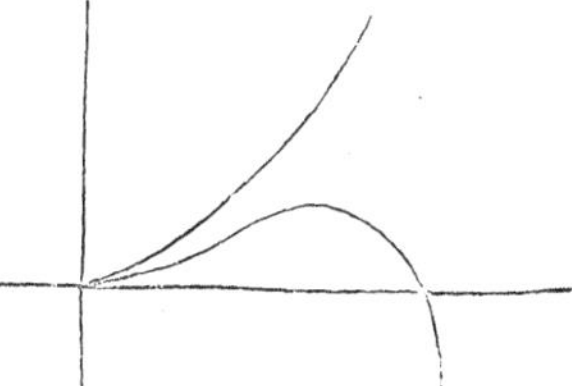

1°. The curve is composed of two branches which extend indefinitely to the right of the origin.

2°. There is a cusp of the second kind at the origin. The tangent is the axis of x.

3°. The lower branch intersects the axis of x at $x = 1$. It is convex to this axis from $x = 0$ to $x = \frac{64}{225}$; thence concave to $x = 1$, from which it becomes convex to the axis of x. The upper branch is always convex to the axis of x.

4°. It has a maximum ordinate to the lower branch for $x = \frac{16}{25}$.

Ex. 4. To trace the curve $y^3 = ax^2 - x^3$.

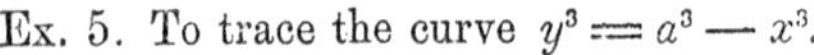

1°. The curve intersects the axis of x at $x = a$, where it has a point of inflexion.

2°. It has a cusp at the origin.

3°. It has an asymptote which makes with the axis of x an angle of 135°.

4°. It has a maximum ordinate for $x = \frac{2}{3}a$.

Ex. 5. To trace the curve $y^3 = a^3 - x^3$.

1°. The curve intersects the axes at $x = a$, $y = a$.

2°. It has an asymptote which passes through the origin.

3°. The points where the curve cuts the axes are points of contrary flexure.

Ex. 6. To trace the curve $y^3 = a^2x - x^3$.

1°. The curve intersects the axis of x at the origin, and also at $x = a$, $x = -a$.

2°. It has an asymptote which passes through the origin.

3°. It has maximum ordinates at $x = \frac{a}{\sqrt{3}}$, $x = -\frac{a}{\sqrt{3}}$.

4°. It has three points of contrary flexure, viz. at the points where it intersects the axes.

5°. It is always concave toward the axis of x.

Ex. 7. To trace the curve $y^2 = ax^2 + bx^3$.

Resolving the equation we shall have $y = \pm (a + bx)^{\frac{1}{2}} x$.

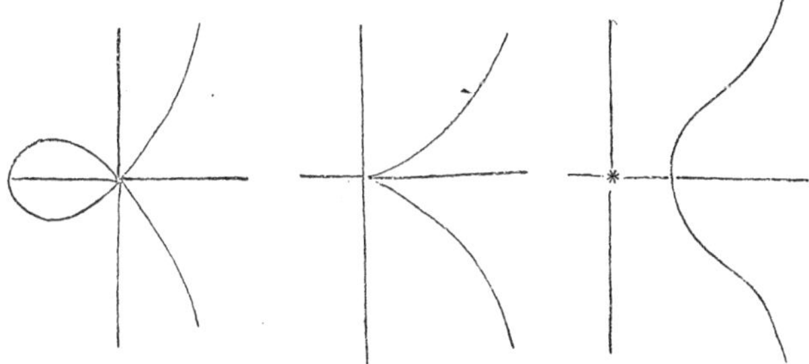

Regarding b always as positive, the curve will take one of the three forms represented in the figures above, according as a is positive, zero, or negative.

V. CONTACT AND CURVATURE OF CURVE LINES.

97. The singular points of a curve and the direction of the curvature between these points enable us to form a very exact idea of the general course of the curve. We proceed, next, to consider the contact of curves, and the method of determining the curvature of a curve line at any point.

Contact and Osculation.

Let there be two curves, referred to the same axes, which have for their equations $y = f(x)$, $y' = f(x')$. Let the abscissas of each receive a common increment $h = \mathrm{PP}'$; then by Taylor's Theorem

$$\mathrm{P}'m = y + \frac{dy}{dx} h + \frac{d^2y}{dx^2} \frac{h^2}{1 \cdot 2} + \frac{d^3y}{dx^3} \frac{h^3}{1 \cdot 2 \cdot 3} + \dots$$

$$\mathrm{P}'m' = y' + \frac{dy'}{dx'} h + \frac{d^2y'}{dx'^2} \frac{h^2}{1 \cdot 2} + \frac{d^3y'}{dx'^3} \frac{h^3}{1 \cdot 2 \cdot 3} + \dots$$

Here, x being made equal to x',

1°. If $y = y' = \text{PM}$, the curves will have a common point of intersection M.

2°. If $y = y'$, and $\frac{dy}{dx} = \frac{dy'}{dx'}$, then the curves have a common tangent, and a contact which is said to be of the *first* order.

3°. In like manner if the curves have a common ordinate, and the first and second differential coefficients in each are equal respectively, they are said to have a contact of the *second* order, and so on.

The tendency of the curves to coincide will, it is evident, increase, as the difference mm' between the ordinates $\text{P}'m'$, $\text{P}'m$ becomes less; or as the number of differential coefficients, which become equal respectively in the two developments, is greater. If all the differential coefficients are equal, the curves will coincide entirely.

98. A line is said to be osculatory to a curve, when it has with the curve a contact of a higher order than any other line of the same species. The osculatory line is called an *osculatrix.*

Ex. As a simple example let it be required to find a right line osculatory to the circle.

Let x and y be the co-ordinates of the right line, x' and y' those of the circle. Then $y = ax + b$, and $x'^2 + y'^2 = \text{R}^2$, will be the equations to the right line and circle respectively. And x being made equal to x', we shall have for the equations of condition

$$y = y', \quad (1) \qquad \frac{dy}{dx} = \frac{dy'}{dx'}. \quad (2)$$

Putting the differential coefficient of the right line equal to that of the circle according to the second condition, we have

$$a = -\frac{x'}{y'}.$$

But by the first condition we have $y' = ax' + b$; whence

$$b = y' + \frac{x'^2}{y'}.$$

Substituting these values of a and b in the equation of the right line $y = ax + b$, and reducing, we obtain

$$xx' + yy' = R^2,$$

which is the equation of a tangent. Thus a tangent line is the osculatrix required.

Osculatory Circle.

99. Let ABC be any curve, T′BT a tangent to the point B, and PB a normal to this point. An infinite number of circles, it is evident, may be described, the centres of which shall be in the normal PB and the circumferences pass through the same point B, and which will have, therefore, a common tangent T′BT. Of these circles, some having a greater curvature than ABC will fall within it, and others having their curvature less than ABC will pass between it and the tangent. There will be, therefore, two classes of tangent circles to the curve; one lying within the curve, and the other without it.

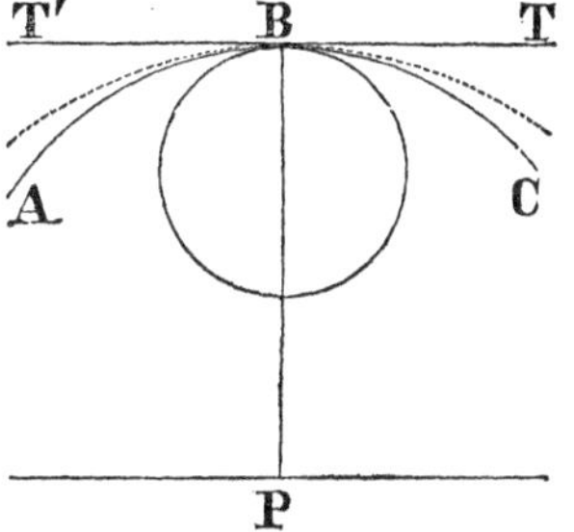

Let it now be required to find among all the possible circles which can be drawn through the point B, that which has the closest contact with the curve ABC, or the *osculatory circle.*

Let x and y be the co-ordinates of the curve ABC, x' and y' those of the circle. Then, supposing the co-ordinates x and y to become equal to x' and y' respectively, we shall have for the equations of condition

$$\frac{dy}{dx} = \frac{dy'}{dx'}, \quad (1) \qquad \frac{d^2y}{dx^2} = \frac{d^2y'}{dx'^2}, \quad (2)$$

or the osculatory circle will be that in which both the differential coefficients are equal respectively to the corresponding ones of the proposed curve. The contact will be of the second order, and the circle is, therefore, called an osculatrix of the *second order.*

100. If, art. 97, the curve Mm be a circle osculatory to the curve Mm', the first three terms in the developments for P$'m$, P$'m'$ will be the same in each. Subtracting the first of these developments from the second, we shall have for h positive

$$P'm' - P'm, \text{ or } mm' = \left(\frac{d^3y'}{dx'^3} - \frac{d^3y}{dx^3}\right)\frac{h^3}{1 \,.\, 2 \,.\, 3} + \&c.;$$

and for h negative

$$mm' = -\left(\frac{d^3y'}{dx'^3} - \frac{d^3y}{dx^3}\right)\frac{h^3}{1 \,.\, 2 \,.\, 3} + \&c.$$

But, h being taken sufficiently small, the first term in each of these series for mm' will exceed the sum of all the terms which follow. Thus mm' will change sign in passing from $x + h$ to $x - h$ and the converse. On one side, therefore, of the point M, the curve will lie above the osculatory circle, while on the other it falls below it. Hence the osculatory circle will, in general, both touch and intersect the curve at the point of osculation. It follows also that the osculatory circle separates the tangent circles which lie without the curve from those which lie within it.

An osculatrix of an even order, we remark in passing, will, in general, intersect as well as touch the curve, while if it be of an odd order it will only touch the curve.

Curvature of Curve Lines.

101. The *curvature* of a curve is its deviation from the tangent. Of two curves, that which departs most rapidly from its tangent has the greatest curvature.

In the same or equal circles the curvature will be the same at all points; but of two different circles, that will depart most rapidly from its tangent, or will have the greatest curvature, the radius of which is the less.

In different circles the curvature is measured by the angle contained between two radii drawn through the extremities of an arc of given length. Thus in two different circles the curvature varies inversely as their radii. The curvature of a curve will, in general, vary from point to point. We have next to determine how the curvature at any point may be found.

102. On account of the closeness of the contact, the curvature at any point of a curve is regarded the same as that of the osculatory circle, and is measured by it. The osculatory circle is, therefore, called the *circle of curvature*, and its radius the *radius of curvature.*

If then we wish to determine the curvature at any point, we have the following problem to solve,—*Given the equation of a plane curve, to find the radius of the osculatory circle, or radius of curvature.*

To resolve the problem, we regard the two curves as having a common ordinate, and make the first and second differential coefficients in each equal respectively. We thus establish relations by means of which, combined with the equation to the circle, the radius of curvature may be found. The same result may be obtained in a more simple manner, if we first find, from the general equation to the circle, an expression for the radius, R, in terms of the first and second differentials, and then substitute for these their values derived from the equation to the proposed curve.

The general equation of the circle, a and b being the co-ordinates of the centre, is

$$(x-a)^2+(y-b)^2=R^2. \qquad (1)$$

Differentiating twice in succession, dx being regarded as constant, we obtain

$$(x-a)\,dx+(y-b)\,dy=0 \qquad (2)$$

$$dx^2+dy^2+(y-b)\,d^2y=0; \qquad (3)$$

whence

$$y-b=-\frac{dx^2+dy^2}{d^2y}, \qquad (4)$$

and

$$x-a=\frac{dy}{dx}\left(\frac{dx^2+dy^2}{d^2y}\right); \qquad (5)$$

substituting these values in (1), we obtain

$$R^2 = \frac{dy^2}{dx^2}\left(\frac{dx^2 + dy^2}{d^2y}\right)^2 + \left(\frac{dx^2 + dy^2}{d^2y}\right)^2,$$

or

$$R^2 = \frac{(dx^2 + dy^2)^3}{(dx\, d^2y)^2};$$

whence

$$R = \pm\frac{(dx^2 + dy^2)^{\frac{3}{2}}}{dx\, d^2y}. \qquad (6)$$

We have thus found an expression for the radius, in terms of the first and second differentials of the co-ordinates. If we now wish to make the circle osculatory to a given curve, we have only to substitute for these differentials their values derived from the equation to the curve; the conditions of osculation will thus be fulfilled, and we shall have an expression for the radius of the osculatory circle, or radius of curvature sought.

The expression for R, which may be regarded as the general formula for the radius of curvature, is affected with the double sign $\pm$, either of which may be taken at pleasure. If we wish the radius to be positive for curves which are concave to the axis of x, we use the negative sign. The radius will then be negative for curves which are convex to the axis.

From the equations (5) and (4) we obtain

$$a = x - \frac{dy}{dx}\left(\frac{dx^2 + dy^2}{d^2y}\right), \qquad (7)$$

$$b = y + \frac{dx^2 + dy^2}{d^2y}, \qquad (8)$$

the general expressions for the co-ordinates of the centre of the osculatory circle. By means of these and the expression for R, equation (6), the osculatory circle to any point of a curve will be completely defined in position and magnitude. And thus the curvature of the curve at any point may be found.

The osculatory circle is not always the curve which approaches most nearly a proposed curve. But on account of the facility with which the circle is described, as well as its readily admitting all degrees of magnitude by the variation of its radius, it is for practical purposes preferred to all other curves having the same order of osculation.

103. Ex. 1. Let it now be proposed to find the radius of curvature of the parabola, whose equation is $y^2=2px$.

Differentiating the equation we obtain

$$dx=\frac{ydy}{p}, \quad dy=\frac{pdx}{y}, \quad d^2y=-\frac{dy^2}{y}.$$

Substituting in the general formula (6) we obtain

$$R=p\left(\frac{y^2}{p^2}+1\right)^{\frac{3}{2}}=\frac{(y^2+p^2)^{\frac{3}{2}}}{p^2}. \qquad (1)$$

Substituting in like manner, in the formulas (7) and (8) the values

$$\frac{dy}{dx}=\frac{p}{y}, \quad dy^2=\frac{p^2dx^2}{y^2}, \quad d^2y=-\frac{p^2dx^2}{y^3},$$

and eliminating y, we find

$$a=3x+p; \quad b=-\frac{2^{\frac{3}{2}}x^{\frac{3}{2}}}{p^{\frac{1}{2}}}, \qquad (2)$$

the formulas for the co-ordinates of the centre of the osculatory circle for any point on the parabola.

As a particular case, let it be required to find the co-ordinates of the centre of the osculatory circle, and the radius of curvature, to a point 9, 6 on a parabola, whose equation is $y^2=4x$.

Ans. $a=29$, $b=-54$, $R=63.245$.

If, in the equation (1) for R, we make $y=0$, we shall have $R=p$. That is, *the radius of curvature at the vertex of the parabola is equal to one half the parameter.*

The expression for the normal to the parabola, art. 87, is $(y^2+p^2)^{\frac{1}{2}}$. We shall have, therefore,

$$R=\frac{(\text{normal})^3}{p^2}.$$

That is, *the radius of curvature at any point of the parabola is equal to the cube of the normal divided by the square of half the parameter ;* hence the radii of curvature at different points are to each other, *as the cubes of the corresponding normals.*

The same principle is true for the conic sections generally.

Ex. 2. To find the radius of curvature to the rectangular hyperbola, $xy = m^2$.

$$\text{Ans. } R = \frac{(x^2+y^2)^{\frac{3}{2}}}{m^2}.$$

VI. INVOLUTES AND EVOLUTES.

104. The centre of the osculatory circle to a curve will have, it is evident, a different position for each point of the curve. And if the several centres of the osculatory circle for each point be joined, the line which joins them will be a curve belonging to the given curve, and deriving its nature and properties from it. This curve, moreover, will have for its co-ordinates the co-ordinates of the centre of the osculatory circle. It is now proposed to find its equation.

The curve, whose equation is sought, is called the *evolute* of the given curve. And, conversely, the given curve compared with its evolute is called the *involute.*

The nature of the evolute is determined, it is evident, by the equations obtained for a and b, the co-ordinates of the centre of the osculatory circle, article 102. If now by means of the equation to the given curve we eliminate from these equations the variable y, and the differentials dy^2, d^2y, &c., we shall obtain, it is evident, two equations in x, from which if x be eliminated, we shall obtain an equation in a, b, and constants only, which will be the equation sought.

Ex. Let the given curve be the common parabola, whose equation is $y^2 = 2px$, to find the equation to the evolute.

Resuming the formulas (2) for a and b, art. 103, and eliminating x, we obtain

$$b^2 = \frac{8}{27p}(a-p)^3;$$

which is the equation to the evolute sought. This equation is that of the semi-cubical parabola.

If we make $b=0$, the equation reduces to $a=p$. Hence the evolute meets the axis of x at a distance from the origin equal to half the parameter.

If we transfer the origin of the co-ordinates to the point, at which the evolute meets the axis of x, we shall have in the equation a for $a-p$; whence by substitution

$$b^2 = \frac{8}{27p}a^3.$$

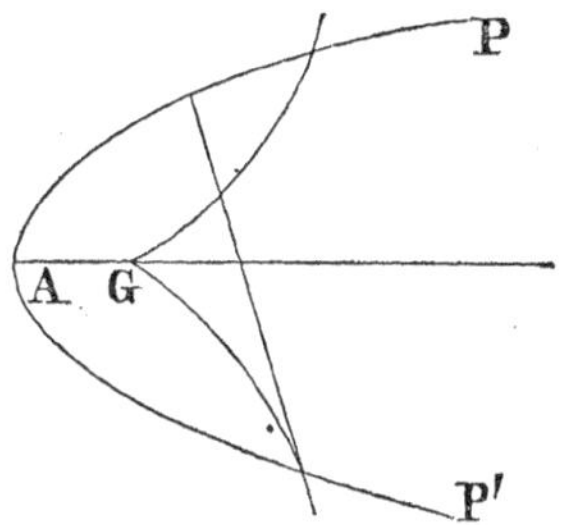

Since the equation gives two values for b with contrary signs, the curve will consist of two branches, symmetrical with respect to the axes, and extending indefinitely in the direction of a positive.

105. We thus see the manner in which the equation of the evolute to any curve may be found. We proceed to investigate some general properties in relation to this curve.

Resuming the first three equations of art. 102, which may be regarded as characterising the osculatory circle, we have

$$(x-a)^2+(y-b)^2=R^2, \qquad (1)$$

$$(x-a)\,dx+(y-b)\,dy=0, \qquad (2)$$

$$dx^2+dy^2+(y-b)\,d^2y=0. \qquad (3)$$

In these equations x and y only have been regarded as variable, a, b, and R being considered constant, as they should be, since our object has been to find the osculatory circle for one point only. But the osculatory circle may be considered as moving along the curve and applying itself to each point in succession. In this case

a, b, and R, it is evident, will also vary with x and y. To determine, therefore, the relations of all these variables, we must differentiate, supposing them all to vary together. Differentiating the equations (1) and (2) on this hypothesis, we obtain

$$(x-a)\,dx+(y-b)\,dy-(x-a)\,da-(y-b)\,db=\mathrm{R}d\mathrm{R},$$

$$dx^2+dy^2+(y-b)\,d^2y-da\,dx-db\,dy=0,$$

which by means of the equations (2) and (3) reduce to

$$-(x-a)\,da-(y-b)\,db=\mathrm{R}d\mathrm{R}, \quad (4)$$

$$-da\,dx-db\,dy=0\,;$$

from which last we obtain $\dfrac{db}{da}=-\dfrac{dx}{dy}$. (5)

But from equation (2) we have

$$y-b=-\frac{dx}{dy}(x-a)\,; \quad (6)$$

which by means of equation (5) becomes

$$y-b=\frac{db}{da}(x-a). \quad (7)$$

These two last, it is evident, are equations for one and the same line. The former is the equation of a normal to the point x, y on the involute, drawn from a point a, b on the evolute; the latter the equation of a tangent to the point a, b on the evolute drawn through the point x, y on the involute. It follows, therefore,

1°. *A normal line to the involute curve is tangent to the evolute.* 2°. The point of tangency on the evolute is *the centre of the osculatory circle.* 3°. The length of the tangent between the involute and the point of tangency on the evolute, *is the radius of the osculatory circle.*

2. Resuming equations (1), (4) and (7) above, we have

$$(x-a)^2+(y-b)^2=\mathrm{R}^2, \quad (1)$$

$$(x-a)\,da+(y-b)\,db=-\mathrm{R}d\mathrm{R}, \quad (2)$$

$$(x-a)\,db-(y-b)\,da=0\,; \quad (3)$$

eliminating $x-a$ from (2) and (3) we obtain

$$\frac{(y-b)(da^2+db^2)}{db}=-\mathrm{R}d\mathrm{R}; \qquad (4)$$

from equation (3) we return to

$$x-a=\frac{da}{db}(y-b); \qquad (5)$$

squaring this last, substituting in (1), and reducing,

$$\frac{da^2+db^2}{db^2}(y-b)^2=\mathrm{R}^2; \qquad (6)$$

next, squaring (4) and dividing the result by (6), member by member,

$$d\mathrm{R}^2=da^2+db^2;$$

whence finally $$d\mathrm{R}=(da^2+db^2)^{\frac{1}{2}};$$

from which it follows that the differential of the radius of curvature is equal to the differential of the arc of the evolute. Hence *The radius of curvature is equal to the arc of the evolute* $\pm$ *a constant.*

Let S represent the arc of the evolute OO′, $C=$ the constant, then we shall have for the radius O′M,

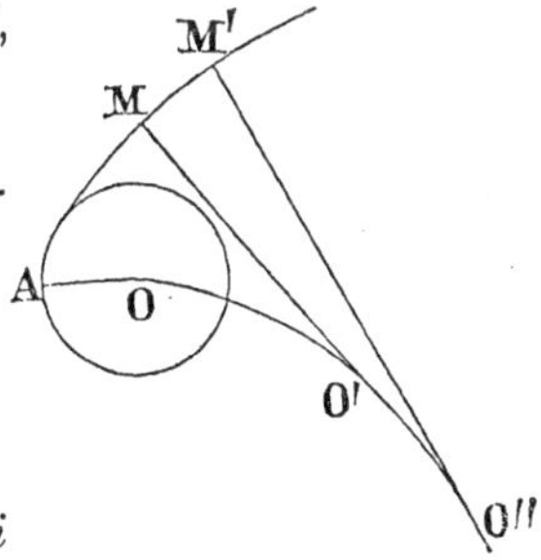

$$\mathrm{R}=\mathrm{S}+C. \qquad (1)$$

In like manner we have for the radius O″M′, $\mathrm{R}'=\mathrm{S}'+C; \qquad (2)$

subtracting the former from the latter

$$\mathrm{R}'-\mathrm{R}=\mathrm{S}'-\mathrm{S}=\mathrm{O}'\mathrm{O}'';$$

hence, *The difference between the radii of curvature at any two points of the involute, is equal to the part of the evolute curve intercepted between them.*

The value of the constant will depend upon the point from which we reckon the arc of the evolute, and may, therefore, be easily determined.

Thus if in the common parabola we reckon the arc from the point at which the evolute meets the axis of x, then, $S = 0$ for $R = p$, and we shall have

$$R = p = 0 + C;$$

whence $C = p$, or the constant is equal to half the parameter.

106. Let a thread be fastened at any point O'' of the evolute, and be applied tightly to the curve. If we now unwind or *evolve* the thread, commencing at A and keeping it always tight, the point A will describe the involute. It is from this relation that the evolute curve has derived its name.

There is some analogy between the manner of conceiving an involute curve to be described and the description of a circle. In the case of the involute the evolute acts as the centre, and the radius is variable.

From what has been done it follows that if the involute is an algebraic curve, the evolute is *rectifiable;* since any arc of it is equal to the difference between the two radii of the osculatory curve drawn from its extremities.

The theory of the evolute, one of the most elegant speculations of geometry, follows, it is evident from what has been done, as a natural sequence to that of the centre and radius of curvature. The honor of its discovery is due to the celebrated Huygens, who derived it directly from the principle we have just seen it involves, viz. the describing a curve by means of a thread evolved or unwrapped from another curve. We shall see hereafter the important application of the theory made by Huygens, in researches in reference to which he was led to its discovery.

We remark in passing that the peculiar artifice of the Calculus is still further seen in the preceding investigations. The set of auxiliary quantities first obtained not being sufficient for our purpose, a new one is introduced, viz. the second differentials, which, alike with the first, are eliminated when the object of their introduction has been effected. The results obtained are thus left, as they should be, in finite and determinate terms.

VII. ENVELOPES TO PLANE LINES.

107. The evolute to a plane curve may be considered as formed by the successive intersections of normals drawn, consecutively, from each point of the curve. From the theory of the evolute we are thus led to seek for some general method for the solution of all problems which depend upon the intersection of lines, whether straight or curved, which vary in position or magnitude according to a given law.

Let $F(x, y, a) = 0$ be the equation to a curve in which a is a parameter to which any number of different values may be assigned. For every particular value assigned to a we shall obtain, it is evident, a particular curve of the class or family represented generally by this equation. Supposing the parameter a, therefore, to vary, let a and $a + da$, be any two successive values of it infinitely near to each other. The two curves which correspond to these values will have for their equations, respectively,

$$F(x, y, a) = 0 \quad (1), \qquad F(x, y, a + da) = 0 \quad (2).$$

The co-ordinates of the points of intersection of the two curves, will be the values of x and y derived from these equations. But the second of these equations returns to

$$F(x, y, a) + d_a F(x, y, a) = 0;$$

a only being supposed to vary. The equations (1) and (2) may, therefore, be replaced by

$$F(x, y, a) = 0 \quad (3), \qquad d_a F(x, y, a) = 0 \quad (4).$$

The point of intersection x, y, determined by these equations, changes its position for every different value assigned to a, and will describe a continuous line when a varies without interruption. If, then, we eliminate a between the equations (3) and (4), the resulting equation, $\varphi(x, y) = 0$, will be the equation to the curve formed by the successive intersections of all the curves derived from the equation $F(x, y, a) = 0$, when a is supposed to vary continuously. The curve, $\phi(x, y) = 0$, possesses the remarkable property that it touches all the curves represented generally by the equation $F(x,y,a) = 0$. On this account it is called the *envelope* to these curves.

The lines represented by the equation $F(x, y, a) = 0$, it should be remarked, may not intersect, however near to each other the consecutive values given to the parameter a, as in the case of the circle $y^2 + x^2 - a = 0$; or the lines may all meet in the same point as in the case of those comprised in the equation $y = ax$. We pass to some examples.

Ex. 1. Let it be required to find the envelope to the series of parabolas whose equation is $y^2 = a(x - a)$, in which a is the parameter.

For the equations of condition we shall have

$$y^2 - a(x - a) = 0 \quad (1), \quad d_a(y^2 - a(x - a)) = 0 \quad (2).$$

Differentiating the equation (2) with reference to a, we obtain $a = \frac{1}{2}x$. Substituting next this value of a in equation (1) we get $y^2 = \frac{1}{4}x^2$, or, $y = \pm \frac{1}{2}x$. The envelope sought will be, therefore, two right lines.

Ex. 2. Find the equation to the curve which touches all the right lines determined by the equation $y = ax + \dfrac{m}{a}$.

Proceeding as above, we find for the equation sought $y^2 = 4mx$. The curve is, therefore, a parabola.

Ex. 3. The straight line PQ slides between two lines AX, AY, at right angles to each other, to find the envelope, or locus of its ultimate intersections.

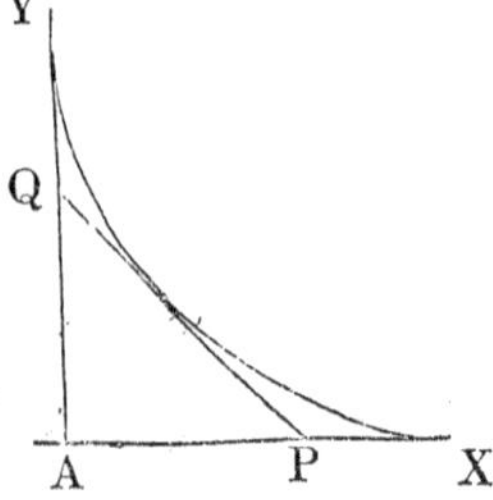

Let AX, AY be taken as the axes of co-ordinates. Let $AP = a$, $AQ = b$, $PQ = c$; then the equation to PQ is

$$\frac{x}{a} + \frac{y}{b} = 1 \quad (1),$$

in which a and b are subject to the condition

$$a^2 + b^2 = c^2 \quad (2).$$

Differentiating these equations with reference to a and b we find

$$\frac{xda}{a^2}+\frac{ydb}{b^2}=0 \ (3); \quad ada+bdb=0 \ (4).$$

By elimination between (3) and (4) we get $\frac{x}{a^3}=\frac{y}{b^3}$, or $\frac{x}{a}=\frac{a^2y}{b^3}$; whence by substitution in (1) $a^2y+b^2y=b^3$, or $b^3=c^2y$.

In like manner we find $a^3=c^2x$. Substituting these values for a and b in (1), we obtain finally for the equation sought

$$x^{\frac{2}{3}}+y^{\frac{2}{3}}=c^{\frac{2}{3}}.$$

Ex. 4. Given the area of a right triangle, to find the curve to which the hypothenuse is always a tangent.

Let the given area equal $2m^2$, AM $=x$, MP $=y$; and let AT $=a$, the parameter by the variation of which TPF changes its position. Then AF $=\frac{4m^2}{a}$, and $y=(a-x)\tan \text{ATF}=(a-x)\frac{4m^2}{a^2}$;

whence $\quad a^2y=4m^2(a-x)$.

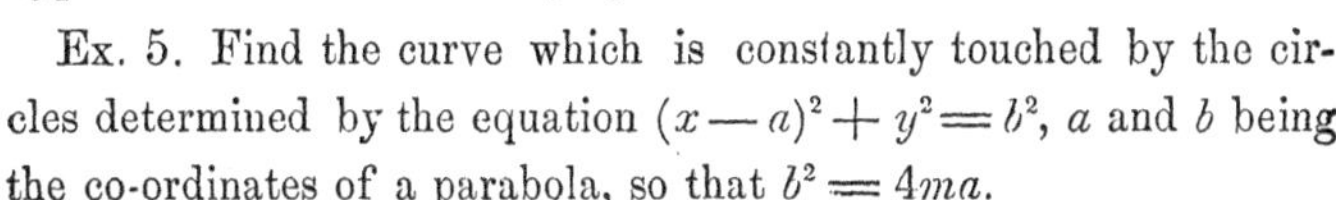

Ans. $xy=m^2$, or the curve is an hyperbola referred to its asymptotes.

Ex. 5. Find the curve which is constantly touched by the circles determined by the equation $(x-a)^2+y^2=b^2$, a and b being the co-ordinates of a parabola, so that $b^2=4ma$.

Ans. $y^2=4m(x+m)$, which is the equation to an equal parabola, having the same axis, and the vertex of which is removed to the distance $-m$.

In this example the circles do not intersect for values of a less than m.

Ex. 6. A projectile is discharged from the point O with a constant velocity, but at different inclinations to the horizon; to find the envelope to the curves it describes.

By the principles of Mechanics the path of the projectile is a parabola represented by the equation

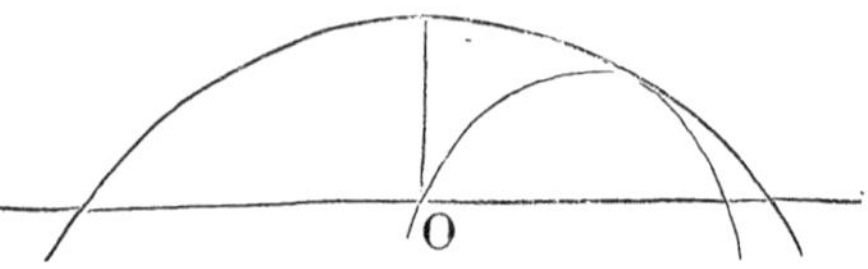

$$y = ax - (1 + a^2)\frac{x^2}{4h};$$

in which x and y are the co-ordinates of the path of the projectile, a the trigonometrical tangent of the angle of projection and h the height due to the velocity of projection.

Ans. $x^2 = 4h(h - y)$ the equation of a parabola having its focus at the point O, and its parameter equal to $4h$. If the surface is required, which bounds all the parabolas that may be described by the body, it will obviously be the parabolid having its focus at O, and parameter equal to $4h$.

This problem was the first of the kind proposed. It was solved by John Bernoulli, but not by any general method. The general process for the solution of all questions of the kind is due to Leibnitz, and is well adapted to exhibit the power and use of the Calculus.

VIII. QUADRATURE. RECTIFICATION OF CURVES.

108. We come next in the theory of curves to the question of their quadrature, rectification and the determination of the surfaces and volumes produced by their revolution. Formulas for this purpose we have already obtained and illustrated with sufficient examples. We have only to remark here that Geometry is indebted to the Integral Calculus for these solutions, while from the Differential Calculus it derives the general method of tangents, the general principles of contact and osculation, and the method of determining the singular points of curves.

IX. DIFFERENTIAL EQUATIONS OF CURVES.

109. From what has been done, it will be seen that by means of differentiation, the constants of an equation may be eliminated, and an equation of a more general character be obtained.

1. Let us take the equation of the right line

$$y = ax + b. \qquad (1)$$

Differentiating and substituting for a its value

$$y = \frac{dy}{dx}x + b\,; \qquad (2)$$

an equation freed from the constant a, and expressing all possible right lines passing through a point determined by the constant b.

Differentiating (1) twice in succession we obtain

$$\frac{d^2y}{dx^2} = 0, \qquad (3)$$

an equation entirely independent of the constants a and b, and which belongs alike to all right lines drawn in the plane of the co-ordinate axes. It is called *the differential equation of lines of the first order.*

2. Let us take next the equation to the circle $y^2 + x^2 = \mathrm{R}^2$.

By differentiation we have $xdx + ydy = 0$;
an equation independent of the radius R, and belonging alike to all circles having their centres at the origin of co-ordinates.

3. Let us take next the general equation of lines of the second order, viz. $y^2 = mx + nx^2$.

Differentiating this equation twice in succession, supposing dx constant, and eliminating m and n from the equations thus obtained and the proposed, we have

$$y^2dx^2 + x^2dy^2 - 2xy\,dxdy + yx^2d^2y = 0,$$

which is *the general differential equation of lines of the second order.*

It is easy to see how we are to proceed in similar cases. The differential equations obtained, it is evident, belong to a species of lines one of which is represented by the proposed equation. They express only the law by which the variable co-ordinates change their values. They ought, therefore, to be independent of the constants, which determine the *magnitude* and not the *nature* of the curve.

SECTION X.

TRANSCENDENTAL FUNCTIONS.

We have thus far investigated rules for the differentiation and integration of Algebraic quantities only. We now proceed to the investigation of those required for the Transcendental Functions.

I. DIFFERENTIATION OF CIRCULAR FUNCTIONS.

110. Let any arc be represented by z, and let it be required to find the differential of $\sin z$.

If we suppose the arc to receive an infinitely small increment dz, it will become $z + dz$, and we shall have for two successive states of the function, $\sin z$, and $\sin (z + dz)$. Whence, taking the difference $$d.\sin z = \sin (z + dz) - \sin z,$$
or, developing $\sin (z + dz)$, Trig. art. 27,

$$d.\sin z = \frac{\sin z \cos dz + \sin dz \cos z}{\mathrm{R}} - \sin z;$$

but, dz being infinitely small, we shall have $\sin dz = dz$, and $\cos dz = \mathrm{R}$; whence
$$d.\sin z = \frac{\cos z\, dz}{\mathrm{R}}.$$

To find, therefore, the differential of the sine of an arc, *Multiply the cosine of the arc by the differential of the arc, and divide the product by the radius.*

2. To find the differential of the cosine of an arc.

By a process altogether similar we obtain

$$d.\cos z = \frac{\cos z \cos dz - \sin z \sin dz}{\mathrm{R}} - \cos z,$$

or reducing $$d.\cos z = -\frac{\sin z dz}{\mathrm{R}}.$$

Thus, to find the differential of the cosine of an arc, *Multiply the sine of the arc taken negatively by the differential of the arc, and divide the product by the radius.*

3. To find the differential of the versed-sine of an arc.

$$d.\ \text{ver-sin}\ z = d.(\text{R} - \cos z) = \frac{\sin z\, dz}{\text{R}}.$$

4. To find the differential of the tangent of an arc.

We have $\quad \tan z = \dfrac{\text{R} \sin z}{\cos z}$; whence, art. 20,

$$d.\ \tan z = \frac{\text{R} \cos z\ d.\sin z - \text{R} \sin z\ d.\cos z}{\cos^2 z};$$

or, by substitution and reduction

$$d.\ \tan z = \frac{(\cos^2 z + \sin^2 z)\, dz}{\cos^2 z} = \frac{\text{R}^2 dz}{\cos^2 z}.$$

Whence, to find the differential of the tangent of an arc, *Multiply the square of the radius by the differential of the arc, and divide the product by the square of the cosine of the arc.*

Since $\quad \cot z = \dfrac{\text{R} \cos z}{\sin z}$, we obtain in like manner

$$d.\ \cot z = -\frac{\text{R}^2 dz}{\sin^2 z}.$$

Whence, to find the differential of the cotangent of an arc, *Multiply the square of the radius taken negatively by the differential of the arc, and divide by the square of the sine of the arc.*

5. Regarding the radius as unity, the preceding formulas become

(1) $d.\ \sin z = \cos z dz$

(2) $d.\ \cos z = - \sin z dz$

(3) $d.\ \text{ver-sin}\ z = \sin z dz$

(4) $d.\ \tan z = \dfrac{dz}{\cos^2 z}$

(5) $d.\ \cot z = - \dfrac{dz}{\sin^2 z}.$

The differentials of the secant and cosecant are occasionally useful. They are readily found, and are as follows

$$(6) \quad d.\sec z = \frac{\sin z dz}{\cos^2 z} = \tan z \sec z dz,$$

$$(7) \quad d.\operatorname{cosec} z = -\frac{\cos z dz}{\sin^2 z} = -\cot z \operatorname{cosec} z dz.$$

111. In the preceding article we have regarded the sine, cosine, &c. as a function of the arc, the arc being regarded as the variable. It is generally most convenient to take the reverse of this, and to regard the arc as the function, and the sine, cosine, &c. as the variable.

To investigate the formulas under this point of view, let z, as before, represent any arc, and let y be its sine; then $y = \sin z$, and $dy = d.\sin z$. We shall have, therefore,

$$dy = \frac{\cos z dz}{\mathrm{R}}; \text{ whence } dz = \frac{\mathrm{R}\, dy}{\cos z},$$

but $\cos z = (\mathrm{R}^2 - \sin^2 z)^{\frac{1}{2}} = (\mathrm{R}^2 - y^2)^{\frac{1}{2}}$; hence

$$dz = \frac{\mathrm{R}\, dy}{(\mathrm{R}^2 - y^2)^{\frac{1}{2}}}, \quad (1)$$

which is the differential of an arc, when its sine is regarded as the independent variable.

2. Let $y = \cos z$; then $dy = d.\cos z$, and we obtain in like manner

$$dz = -\frac{\mathrm{R} dy}{\sin z} = -\frac{\mathrm{R}\, dy}{(\mathrm{R}^2 - y^2)^{\frac{1}{2}}}, \quad (2)$$

which is the differential of an arc, when its cosine is regarded as the independent variable.

3. Let $y = \text{ver-sin}\, z$; then $dy = d.\text{ver-sin}\, z$; and we have

$$dz = \frac{\mathrm{R} dy}{\sin z};$$

but $\cos z = \mathrm{R} - y$, and $\sin z = [\mathrm{R}^2 - (\mathrm{R} - y)^2]^{\frac{1}{2}} = (2\mathrm{R}y - y^2)^{\frac{1}{2}}$;

whence

$$dz = \frac{\mathrm{R}\, dy}{(2\mathrm{R}y - y^2)^{\frac{1}{2}}}, \quad (3)$$

which is the differential of an arc, its versed-sine being regarded as the independent variable.

4. Let $y = \tan z$; then $dy = d.\tan z$; whence

$$dz = \frac{\cos^2 z\,dy}{R^2};$$

but $\frac{\cos z}{R} = \frac{R}{\sec z}$; hence $\frac{\cos^2 z}{R^2} = \frac{R^2}{\sec^2 z} = \frac{R^2}{R^2 + \tan^2 z}$,

whence
$$dz = \frac{R^2 dy}{R^2 + y^2}, \qquad (4)$$

the differential of an arc, its tangent being the independent variable.

112. The following notation has recently been introduced into the Calculus, and is found of great convenience.

$\text{Sin}^{-1} y =$ the arc whose sine is y,

$\text{Cos}^{-1} y =$ the arc whose cosine is y, &c. &c.

Employing this notation the four preceding formulas, together with those for the cotangent, secant and cosecant, will be expressed as in the following table.

	$R = 1$	$R = a$
1.	$d.\sin^{-1} y = \frac{dy}{(1 - y^2)^{\frac{1}{2}}}$	$d.\sin^{-1} y = \frac{a\,dy}{(a^2 - y^2)^{\frac{1}{2}}}$
2.	$d.\cos^{-1} y = -\frac{dy}{(1 - y^2)^{\frac{1}{2}}}$	$d.\cos^{-1} y = -\frac{a\,dy}{(a^2 - y^2)^{\frac{1}{2}}}$
3.	$d.\text{ver-sin}^{-1} y = \frac{dy}{(2y - y^2)^{\frac{1}{2}}}$	$d.\text{ver-sin}^{-1} y = \frac{a\,dy}{(2ay - y^2)^{\frac{1}{2}}}$
4.	$d.\tan^{-1} y = \frac{dy}{1 + y^2}$	$d.\tan^{-1} y = \frac{a^2 dy}{a^2 + y^2}$
5.	$d.\cot^{-1} y = -\frac{dy}{1 + y^2}$	$d.\cot^{-1} y = -\frac{a^2 dy}{a^2 + y^2}$
6.	$d.\sec^{-1} y = \frac{dy}{y(y^2 - 1)^{\frac{1}{2}}}$	$d.\sec^{-1} y = \frac{a^2 dy}{y(y^2 - a^2)^{\frac{1}{2}}}$
7.	$d.\text{cosec}^{-1} y = -\frac{dy}{y(y^2 - 1)^{\frac{1}{2}}}$	$d.\text{cosec}^{-1} y = -\frac{a^2 dy}{y(y^2 - a^2)^{\frac{1}{2}}}$

We pass next to some examples, regarding R as 1.

Ex. 1. Let $u = x\sin^{-1}x$, to find the differential of u.

$$\text{Ans.}\quad du = \sin^{-1}x dx + \frac{x dx}{(1-x^2)^{\frac{1}{2}}}.$$

Ex. 2. Let $u = (\sin^{-1}x)^2$, to find du.

$$\text{Ans.}\quad 2\sin^{-1}x\,\frac{dx}{(1-x^2)^{\frac{1}{2}}}.$$

Ex. 3. Let $u = \tan^{-1}\frac{x}{2}$, to find du. Ans. $\frac{2dx}{4+x^2}$.

Ex. 4. Given $u = \tan^{-1}\frac{x}{y}$, to find du.

Let $\frac{x}{y} = t$, then $du = \frac{dt}{1+t^2}$, and hence

$$du = \frac{ydx - xdy}{y^2 + x^2}.\quad \text{Ans.}$$

113. In the Trigonometry we have found rules by which to compute the sines, cosines, &c. for every minute in a quadrant. The Calculus enables us to develop new and more convenient theorems for this object.

Resuming the Theorem of Maclaurin,

$$y = (y) + \left(\frac{dy}{dx}\right)x + \frac{1}{2}\left(\frac{d^2y}{dx^2}\right)x^2 + \frac{1}{2\,.\,3}\left(\frac{d^3y}{dx^3}\right)x^3 + \dots$$

let it be proposed to develop the sine and cosine of an arc in terms of the arc itself.

Let $y = f(x) = \sin x$; then, radius being unity,

$$\frac{du}{dx} = \cos x,\quad \frac{d^2u}{dx^2} = -\sin x,\quad \frac{d^3u}{dx^3} = -\cos x,\ \&c.$$

whence, putting $x = 0$, these values become

$$\frac{du}{dx} = 1,\quad \frac{d^2u}{dx^2} = 0,\quad \frac{d^3u}{dx^3} = -1,\ \&c.;$$

and we have $$\sin x = \frac{x}{1} - \frac{x^3}{1.2.3} + \frac{x^5}{1.2.3.4.5} - \&c.$$

2. To develop the cosine in terms of the arc.

Proceeding in a manner altogether similar, we obtain

$$\cos x = 1 - \frac{x^2}{1.2} + \frac{x^4}{1.2.3.4} - \&c.$$

The two series converge rapidly when the arc is small and are of great utility in the calculation of the trigonometrical tables.

DIFFERENTIATION OF EXPONENTIAL FUNCTIONS.

114. Let it be proposed next to find the differential of the quantity a^x, a being constant and x a variable exponent. For two successive states of the quantity we have a^x and a^{x+dx}; whence

$$d\,a^x = a^{x+dx} - a^x,$$

or

$$d\,a^x = a^x\,(a^{dx} - 1). \quad (1)$$

But to render this expression of service we must develop the term a^{dx}, so that dx shall not be employed in the expression as an exponent. In order to this we assume $a = 1 + b$, and developing by the binomial formula, we have

$$a_{dx} = (1+b)^{dx} = 1 + dx b + dx\frac{dx-1}{1.2}b^2 + dx\frac{(dx-1)(dx-2)}{1.2.3}b^3 +,$$

whence, since dx is infinitely small by the side of the numbers 1, 2, &c., we have by transposition and reduction

$$a^{dx} - 1 = dx\left(b - \frac{b^2}{2} + \frac{b^3}{3} - \frac{b^4}{4} + \ldots\right); \quad (2)$$

hence, by substitution in (1) and replacing b by its value $a - 1$,

$$d\,a^x = a^x dx\,[(a-1) - \tfrac{1}{2}(a-1)^2 + \tfrac{1}{3}(a-1)^3 - \ldots]. \quad (3)$$

115. Let it now be proposed to find the differential of the logarithm of a quantity.

Whatever the base a of a system of logarithms, we shall have, x being any number,

$$x = \log a^x;\ \text{whence}\ dx = d.\log a^x;$$

substituting this value of dx in equation (3)

$$d\,a^x = a^x d.\log a^x[(a-1) - \tfrac{1}{2}(a-1)^2 + \tfrac{1}{3}(a-1)^3 - \&c.]$$

Let $a^x = y$; then we shall have

$$d. \log y = \frac{dy}{y} \frac{1}{[(a-1)-\frac{1}{2}(a-1)^2+\frac{1}{3}(a-1)^3 \&c.]}.$$

Representing this last factor, which is constant, by M, we shall have for any system of logarithms

$$d. \log y = \frac{Mdy}{y}.$$

The fraction M, which depends upon the base a of the system of logarithms employed, is called the *modulus* of the system. Hence in any system, *the differential of the logarithm of a quantity is equal to the modulus of the system, into the differential of the quantity divided by the quantity itself.*

If we take the most simple case, and suppose $M = 1$, the system will be the Naperian, and we shall have

$$d. \log y = \frac{dy}{y}.$$

Hence, *the differential of the Naperian logarithm of a quantity is equal to the differential of the quantity divided by the quantity itself.*

The Naperian logarithms, so called from the name of their inventor, are those usually employed in algebraic calculations. They are usually designated by l' or $\log_e$. We shall employ the common designation *log.* to express them. The Naperian logarithms will be understood, therefore, by this term, unless the contrary is expressed.

If from the expression for M we calculate the value of a for the hypothesis $M = 1$, it will be 2.71828 nearly. This number is the base of the Naperian system of logarithms. It is of frequent use, and is commonly designated by e. Thus we have

$$e = 2.71828.$$

We pass to some examples.

Ex. 1. Find the differential of $\log(a+x)$. Ans. $\frac{dx}{a+x}$.

Ex. 2. Find the differential of $\log \frac{a}{a+x}$.

$$d. \log \frac{a}{a+x} = d[\log a - \log(a+x)]. \quad \text{Ans. } -\frac{dx}{a+x}.$$

Ex. 3. Find the differential of $\log(xy)$. Ans. $\frac{dx}{x}+\frac{dy}{y}$.

Ex. 4. Find the differential of $\log\frac{x}{y}$. Ans. $\frac{dx}{x}-\frac{dy}{y}$.

Ex. 5. Find the differential of $\log\frac{a+x}{a-x}$.

Ans. $\frac{dx}{a+x}+\frac{dx}{a-x}$.

Ex. 6. Find the differential of $\log(a^2+x^2)^{\frac{1}{2}}$.

$$d. \log(a^2+x^2)^{\frac{1}{2}} = d.\tfrac{1}{2}\log(a^2+x^2). \quad \text{Ans. } \frac{xdx}{a^2+x^2}.$$

Ex. 7. Find the differential of $\log[x+(1+x^2)^{\frac{1}{2}}]$.

Ans. $\frac{dx}{(1+x^2)^{\frac{1}{2}}}$.

116. The differentials of exponential functions are readily found by the rule for the differentiation of logarithmic functions.

1. Let the function be $u = a^x$.

We have $\log u = x \log a$;

whence $d. \log u = \log a dx$,

or $\frac{du}{u} = \log a dx$;

whence $du = \log a\, a^x dx$.

Thus the differential of an exponential function, whose exponent only is variable, is equal to the continued product of the Naperian logarithm of the base, the function itself and the differential of its exponent.

2. Let the function be $u = x^y$.

We have $$\log u = y \log x;$$

whence $$\frac{du}{u} = dy \log x + y \frac{dx}{x},$$

and $$du = x^y \log x dy + yx^{y-1}dx.$$

Thus the differential of an exponential function when both the base and the exponent are variable, is equal to the sum of the differentials found by considering each separately as constant and the other as variable.

INTEGRATION OF CIRCULAR FUNCTIONS.

117. Reversing the process of differentiation, we shall have for the integrals of the four fundamental elementary differentials, art. 112, radius being 1,

$$(1) \int \frac{dy}{(1-y^2)^{\frac{1}{2}}} = \sin^{-1}y, \quad (3) \int \frac{dy}{1+y^2} = \tan^{-1}y,$$

$$(2) \int \frac{dy}{(2y-y^2)^{\frac{1}{2}}} = \text{ver-sin}^{-1}y, \quad (4) \int \frac{dy}{y(y^2-1)^{\frac{1}{2}}} = \sec^{-1}y,$$

to each of which, in the practical application of them, a constant C should be added.

In order to see how the constant is determined, we take the formula (1) for example.

$$\int \frac{dy}{(1-y^2)^{\frac{1}{2}}} = \sin^{-1}y + C.$$

Here, if we estimate the arc from the beginning of the first quadrant, since the sine will be 0 when the arc is 0, we shall have, it is evident, $C = 0$.

Let the arc be estimated next from the beginning of the second quadrant. This supposition will render

$$\int \frac{dy}{(1-y^2)^{\frac{1}{2}}} = 0, \text{ for } y = 1.$$

But $y = 1$ gives $\sin^{-1}y = \frac{1}{2}\pi$; whence $0 = \frac{1}{2}\pi + C$, or $C = -\frac{1}{2}\pi$, and we have for the corrected integral,

$$\int \frac{dy}{(1-y^2)^{\frac{1}{2}}} = \sin^{-1}y - \tfrac{1}{2}\pi.$$

118. Expressions which do not appear in the above forms may often be readily reduced to them.

Thus let it be required to find the integral of

$$du = \frac{dx}{(a^2 - x^2)^{\frac{1}{2}}}.$$

Put $x = av$, then $dx = adv$, and $(a^2 - x^2)^{\frac{1}{2}} = a(1 - v^2)^{\frac{1}{2}}$;

whence by substitution $\quad du = \dfrac{dv}{(1 - v^2)^{\frac{1}{2}}}$,

and we have, comparing with formula (1) above,

$$u = \int \frac{dv}{(1 - v^2)^{\frac{1}{2}}} = \sin^{-1}v;$$

whence, replacing v by its value,

$$\int \frac{dx}{(a^2 - x^2)^{\frac{1}{2}}} = \sin^{-1}\frac{x}{a}. \qquad (1)$$

In a manner altogether similar the following integrals are obtained.

$$\int \frac{dx}{(2ax - x^2)^{\frac{1}{2}}} = \text{ver-sin}^{-1}\frac{x}{a} \qquad (2)$$

$$\int \frac{dx}{a^2 + x^2} = \frac{1}{a}\tan^{-1}\frac{x}{a} \qquad (3)$$

$$\int \frac{dx}{x(x^2 - a^2)^{\frac{1}{2}}} = \frac{1}{a}\sec^{-1}\frac{x}{a}. \qquad (4)$$

The formulas above are called *circular forms.*

INTEGRATION OF LOGARITHMIC FUNCTIONS.

119. Reversing the process of differentiation, art. 115, we have

$$\int \frac{dx}{x} = \log x.$$

In like manner $\int \frac{dx}{(a+x)} = \log (a + x)$.

In general, *if the numerator of a fraction is the differential of the denominator, the integral will be equal to the Naperian logarithm of the denominator.*

Ex. 1. Find the integral of $\frac{2bxdx}{a+bx^2}$. Ans. $\log (a + bx^2)$.

Ex. 2. Find the integral of $\frac{2xdx}{1+x^2}$. Ans. $\log (1 + x^2)$.

Ex. 3. Find the integral of $\frac{adx}{1+x}$. Ans. $a \log (1 + x)$.

Ex. 4. Find the integral of $\frac{dx}{bx}$. Ans. $\frac{1}{b} \log x$.

In general if the numerator of a fraction is the differential of the denominator *multiplied or divided by a constant quantity*, the integral will be the Naperian logarithm of the denominator *multiplied or divided by the same constant quantity.*

120. Expressions which do not appear in the logarithmic forms may frequently be reduced to them.

1. To find the integral of $\frac{dx}{(x^2-a^2)^{\frac{1}{2}}}$.

Put $x^2 - a^2 = z^2$, then $\frac{dz}{dx} = \frac{x}{z}$, and we have

$$dz + dx : dz :: x + z : x;$$

or

$$\frac{dx}{z} = \frac{dz+dx}{x+z};$$

whence

$$\int \frac{dx}{(x^2-a^2)^{\frac{1}{2}}} = \int \frac{dx}{z} = \log (x + z);$$

or replacing z by its value, and introducing the double sign, since the processes in the two cases will be altogether the same, we have

$$\int \frac{dx}{(x^2 \pm a^2)^{\frac{1}{2}}} = \log [x + (x^2 \pm a^2)^{\frac{1}{2}}]. \qquad (1)$$

2. To find the integral of $\frac{dx}{x(a^2+x^2)^{\frac{1}{2}}}$.

Put $x = \frac{a}{z}$, then $\frac{dx}{x(a^2+x^2)^{\frac{1}{2}}} = -\frac{dz}{a(z^2+1)^{\frac{1}{2}}} = -\frac{1}{a}\frac{dz}{(z^2+1)^{\frac{1}{2}}}$;

which reduces to the preceding case, and we have

$$\int \frac{dx}{x(a^2+x^2)^{\frac{1}{2}}} = -\frac{1}{a}\int \frac{dz}{(z^2+1)^{\frac{1}{2}}} = -\frac{1}{a}\log [z+(z^2+1)^{\frac{1}{2}}].$$

Replacing z by its value, and using the double sign as before, we shall have

$$\int \frac{dx}{x(a^2 \pm x^2)^{\frac{1}{2}}} = -\frac{1}{a}\log \left(\frac{a+(a^2 \pm x^2)^{\frac{1}{2}}}{x}\right). \qquad (2)$$

3. To find the integral of $\frac{dx}{a^2 - x^2}$.

Recollecting that $\frac{2a}{a^2-x^2} = \frac{2a}{(a+x)(a-x)} = \frac{1}{a+x} + \frac{1}{a-x}$,

we obtain $$\int \frac{dx}{a^2 - x^2} = \frac{1}{2a}\log \left(\frac{a+x}{a-x}\right). \qquad (3)$$

Ex. 4. To find the integral of $\frac{dx}{x^2-a^2}$.

By a process altogether similar to that pursued in the last case, we obtain

$$\int \frac{dx}{x^2 - a^2} = \frac{1}{2a}\log \left(\frac{x-a}{x+a}\right). \qquad (4)$$

The formulas thus obtained are called *logarithmic forms.*

The circular and logarithmic forms, together with the more simple algebraic form $\int x^m dx = \frac{x^{m+1}}{m+1}$, constitute the fundamental formulas. To one or more of these forms it is the object of almost every process of the Integral Calculus to reduce the integrals of proposed algebraic expressions. Before proceeding further, however, we shall make some application of the additional principles now obtained.

SECTION XI.

APPLICATION OF THE CALCULUS TO TRANSCENDENTAL CURVES.

121. Curves may be divided into two general classes.

1°. Those whose equations are purely algebraic. These are called *Algebraic Curves.*

2°. Those whose equations involve transcendental expressions. These are called *Transcendental Curves.*

We have already considered those of the first class. We will now take some examples of the second.

I. THE SINUSOID.

122. A curve formed by taking for abscissas the arcs of a circle from 0 to any number of circumferences, and for the ordinates the corresponding sines of these arcs, is called the *curve of sines*, or, the *sinusoid.*

1. *Figure of the curve.* The equation to the curve is $y = \sin x$, and for the first and second differential coefficients we have, r being the radius,

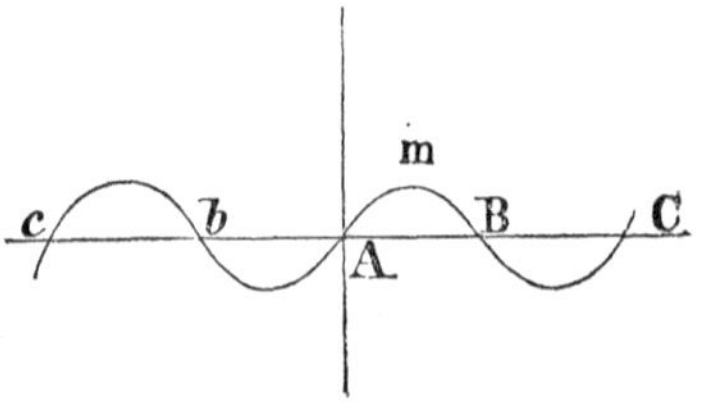

$$\frac{dy}{dx} = \frac{\cos x}{r}, \quad \frac{d^2y}{dx^2} = -\frac{\sin x}{r^2};$$

and by the process explained art. 96, we find

1°. The curve is always concave toward the axis of x.

2°. It intersects the axis at points for which we have $x = 0$, $x = \pi r$, $x = 2\pi r$, &c.

3°. For all values of x from $x=0$ to $x=\pi r$, y is positive, and the curve lies above the axis of x; from $x=\pi r$ to $x=2\pi r$ it is negative, and the curve lies below the axis of x; and so on alternately.

4°. It has maximum ordinates, $y=+r$, $y=-r$, for $x=\frac{1}{2}\pi r$, $x=\frac{5}{2}\pi r$, &c., and $x=\frac{3}{2}\pi r$, $x=\frac{7}{2}\pi r$, &c.

5°. It has points of inflexion at $x=\pi r$, $2\pi r$, &c.

6°. The tangents through the points of inflexion intersect the axis at an angle of 45°.

2. *Area of the curve.* For the element of the area, we have

$$ydx = \sin xdx;$$

whence
$$A = \int \sin xdx;$$

and integrating to radius r,

$$A = -r\cos x + C.$$

Taking the integral between $x=0$, $x=\pi r$, we have

$$A = 2r^2;$$

whence, the area of the surface AmB is equal to *twice the square of the radius.*

II. THE LOGARITHMIC CURVE.

123. A curve referred to rectangular co-ordinates, in which one of the co-ordinates is the logarithm of the other, is called the *logarithmic curve.*

Let a represent the base of a system of logarithms; then $a^x=y$ will be the equation to the curve. And since different values may be assigned to a, there will be as many different logarithmic curves as there are different values assigned to a.

If we make $x=0$, we shall have $y=1$; and as we have the same result whatever the value of a, it follows that *every logarithmic curve will intersect the axis of numbers at a distance from the origin equal to unity.*

1. *Figure of the curve.* If we make $x=0$, $x=\frac{1}{2}$, $x=1$, $x=\frac{3}{2}$, &c. we shall have for the corresponding values of y, $y=1$, $y=\sqrt{a}$, $y=a$, $y=a\sqrt{a}$, &c.

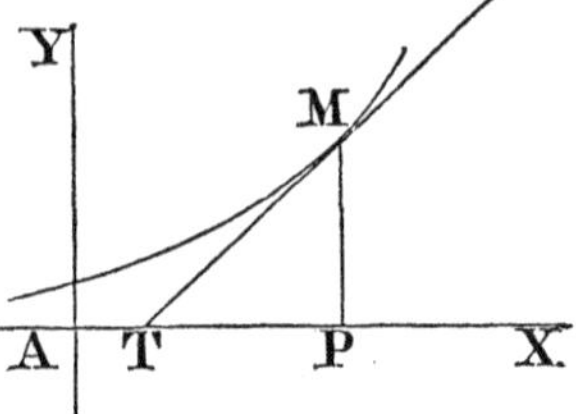

Setting off the values of x upon the axis AX and erecting the corresponding ordinates, the curve may be constructed by a series of points.

If we suppose $a>1$, the logarithms of numbers greater than unity being positive, the values of x will be set off to the right of the origin A. And the logarithms of numbers less than unity being negative, the corresponding values of x will be set off to the left of the origin.

For $y=0$, $x=-\infty$. Thus the axis AX is an asymptote to the curve in the direction of x negative. For $y=\infty$, $x=\infty$. The curve extends indefinitely to the right of the origin above the axis of x.

2. *Subtangent to the curve.*

Differentiating the equation of the curve, we have, art. 115, M being the modulus,

$$dx=\frac{Mdy}{y}.$$

Substituting the value of $\frac{dx}{dy}$ in the general expression for the subtangent, art. 30, we obtain

$$\text{PT}=M.$$

That is, the subtangent to the logarithmic curve is constant, and *is equal to the modulus of the system of logarithms from which the curve is constructed.* This is the most remarkable property of the curve.

In the Naperian system $M=1$; hence $\text{PT}=1$.

3. *Area of the curve.* To find the area of the curve we have by substitution in the element of the area and integrating,

$$\text{A}=My+\text{C}.$$

Taking the integral between the limits y' and y we have

$$A = M(y - y'),$$

that is, *the area comprised between any two ordinates is equal to the rectangle under the subtangent and the difference between the ordinates.*

If we suppose $y' = 0$, $A = My$, that is, the area extending from any ordinate indefinitely on the side on which the curve approaches its asymptote, *is equal to the ordinate multiplied by the subtangent.*

Thus the area extending indefinitely to the left of PM is equal to PM multiplied by PT.

4. *Solid of revolution formed by the curve.*

If the curve revolve about the axis AX, we shall have for the element dV of the solid thus generated,

$$dV = M\pi y dy,$$

whence $$V = \tfrac{1}{2}\pi y^2 \times M.$$

Thus the solid produced is equal to *once and a half the cone generated at the same time by the triangle having* y *for its base and M for its altitude.*

The leading properties of this curve were demonstrated by Huygens by the laborious processes of the Ancient Geometry. We see with what facility they are obtained by the Calculus. The curve itself is moreover not without interest. It admits of important applications in Physical Science. Thus, since the density of the atmosphere decreases geometrically as its altitude increases arithmetically, its density may be represented by a logarithmic curve, the altitudes being measured upon the asymptote.

POLAR CO-ORDINATES OF CURVES.

124. The curves heretofore considered have been determined by rectangular co-ordinates.

But a point in a curve may also be determined by the radius vector AM and the angle MAP which this radius makes with a fixed axis AX.

The angle MAP is measured by the arc *om* to the radius 1. The arc *om* may, therefore, be substituted for this angle. The radius vector AM and the arc *om* will then be the *polar co-ordinates* of the curve. The point *o* may be regarded as the origin.

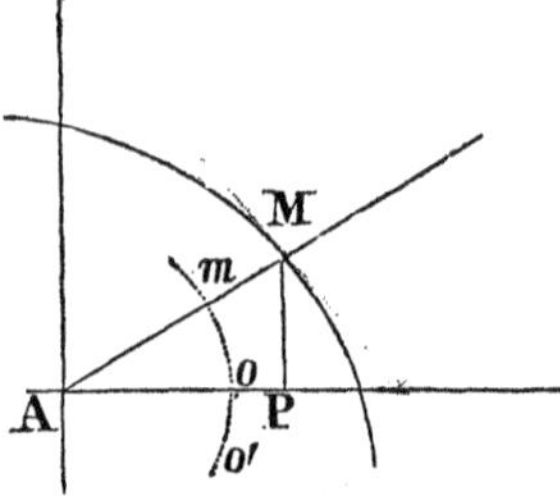

It will be easy to pass from a system of rectangular co-ordinates to the polar co-ordinates of a curve.

Let $AM = r$, $om = t$, $AP = x$, $PM = y$. To determine the relation which exists between these co-ordinates we have

$$AP = AM \cos MAP, \qquad PM = AM \sin MAP,$$

or

$$x = r \cos t, \qquad y = r \sin t.$$

If we now substitute these values of x and y in the equation to a curve expressed in terms of rectangular co-ordinates, we shall have the equation to the curve in polar co-ordinates, or the *polar equation* of the curve.

If the origin of the rectangular co-ordinates x and y is taken at a point A′ different from A, let x', y' be the co-ordinates reckoned from A′, and a, b the co-ordinates of A reckoned also from A′. Then we shall have

$$x = x' - a, \quad y = y' - b,$$

which must be substituted for x and y in the expressions above obtained.

Subtangent to Polar Curves.

125. In a curve expressed by rectangular co-ordinates the subtangent is the line comprised between the foot P of the ordinate and the point T where the perpendicular to the ordinate, drawn through P, intersects the tangent. In polar curves the ordinate is the radius vector AM. Retaining, then, the same definition, the subtangent to a point M in the polar curve will be the line AT (fig. below) comprised between A the foot of the ordinate and the

point T where the perpendicular to the ordinate, drawn through A, intersects the tangent MT.

The subtangent in polar curves is, therefore, different from what it is in curves defined by rectangular co-ordinates. For in the latter it is always reckoned on the axis of abscissas; but this axis does not exist in polar curves, and the subtangent varies its position at each point of the curve.

To determine the subtangent AT, let M′ be a point on the curve infinitely near to the point M, and AM′ the corresponding radius vector. With the radius AM describe from A, as a centre, the arc MQ meeting AM′ in Q, and, with a radius $= 1$, describe from A an arc mq meeting AM and AM′ in m and q. Let $t =$ the measuring arc, and r the radius vector AM. Then, since the angle M′AM is infinitely small, $mq = dt$, and $\text{QM}' = dr$. The arc QM, moreover, may be considered as a right line perpendicular to AM, and the triangle M′QM will be similar to the triangle MAT; and we shall have

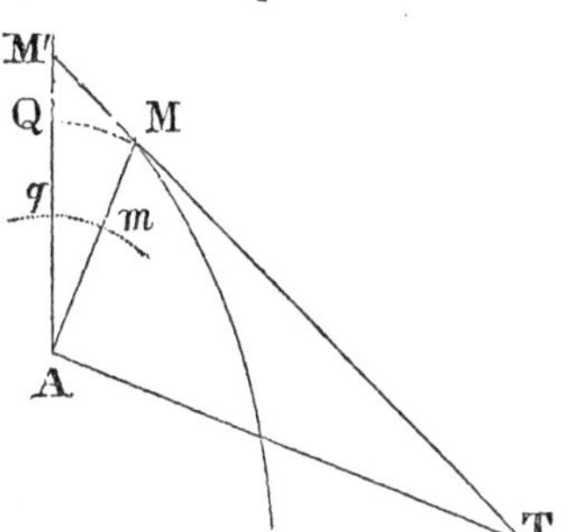

$$\text{AT} : \text{AM} :: \text{QM} : \text{QM}',$$

or

$$\text{AT} : r :: \text{QM} : dr.$$

But, from the similar sectors $m\text{A}q$, QAM, we find $\text{QM} = rdt$; whence by substitution and reduction we obtain

$$\text{AT} = \frac{r^2 dt}{dr}, \qquad (1)$$

the general expression for the subtangent to the polar curve.

Differential of the arc, and of the surface of Polar Curves.

126. Let $z =$ the infinitely small arc MM′ in the preceding figure. Then from the triangle MQM′, right angled at Q, we have

$$\text{MM}' = (\text{MQ}^2 + \text{QM}'^2)^{\frac{1}{2}};$$

or substituting for these lines their values, as above, we obtain

$$dz = (r^2dt^2 + dr^2)^{\frac{1}{2}}, \qquad (2)$$

the differential of any arc of a polar curve.

2. For the differential of the surface we shall have, it is evident, the infinitely small triangle $\text{AMM}' = \frac{1}{2}\text{AM}' \times \text{QM}$. Substituting for these lines as above, we have

$$d\text{A} = \tfrac{1}{2}r^2dt, \qquad (3)$$

the expression for the differential of the surface of a polar curve.

POLAR CURVES. SPIRALS.

127. A spiral is a curve described by a point which moves along a right line in accordance with some law, the right line having at the same time a uniform angular motion.

Thus let PD be a straight line which moves with a uniform angular motion about the point P. When the motion begins, let a point M move also along the line PD according to any prescribed law; the point M will describe a spiral.

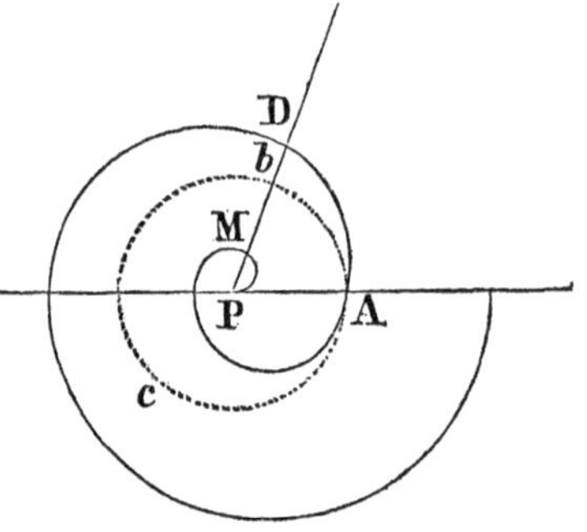

The fixed point P about which the right line moves is called the pole of the spiral; the point M is called the *generating* point. The portion of the spiral generated in each revolution is called a *spire.*

Let PA be the radius vector at the end of the first revolution. With P as a centre, and PA as radius, describe a circle A*bc*. The angular motion of the radius vector may be measured by the arcs of this circle estimated from A. A*bc* is, therefore, called the *measuring circle.*

Referring to the above figure, we proceed to some of the spirals which have been an object of particular attention to geometricians.

Spiral of Archimedes.

128. If when the line PD moves uniformly about the pole P, the point M moves also uniformly along the line PD, the spiral is called *the spiral of Archimedes.*

To find its equation, let the radius vector PM $= r$, the measuring arc Ab $= t$, and the radius PA of the measuring circle $= a$. Then, since the radius vectors are proportionate to the measuring arcs, we have

$$\text{PM} : \text{PA} :: \text{arc A}b : \text{circ A}bc,$$

or

$$r : a :: t : 2\pi a,$$

whence

$$r = \frac{t}{2\pi},$$

which is the polar equation sought.

1. To find the subtangent.

Substituting in the general expression for the subtangent, AT $= \frac{r^2dt}{dr}$, we obtain, P being the pole,

$$\text{PT} = \frac{t^2}{2\pi},$$

the expression for the subtangent sought.

If we make $t = 2\pi$, we shall have AT $= 2\pi$. Hence, *the subtangent after the first revolution, is equal to the circumference of the measuring circle.*

If we suppose m revolutions we shall have $t = 2m\pi$, whence

$$\text{PT} = 2m^2\pi = m.2m\pi;$$

hence, *the subtangent after* m *revolutions is equal to* m *times the circumference of the circle described with the radius vector to the point of contact.*

2. To find the area.

Substituting in the general element for the surface of a polar curve, $dA = \frac{1}{2}r^2dt$, we obtain for the element of the spiral

$$dA = \frac{t^2dt}{8\pi^2}.$$

Integrating between the limits $t=0$, $t=2\pi$, we obtain $A=\frac{1}{3}\pi$, the area described by one revolution of the radius vector. Thus, *the area included in the first spire is equal to one third the area of a circle, whose radius is equal to the radius vector after the first revolution.*

If we integrate between the limits $t=0$, $t=2.2\pi$, we have $A=\frac{8}{3}\pi$, the whole area described by the radius vector during two revolutions. But since the area described in the first revolution is repeated in the second, we shall have for the actual spiral area at the end of the second revolution

$$A=\tfrac{8}{3}\pi-\tfrac{1}{3}\pi=\tfrac{7}{3}\pi.$$

This spiral was invented by Conon, a friend of Archimedes. Its properties, however, were discovered after laborious research by Archimedes, and hence its name.

Hyperbolic Spiral.

129. If while the line PB revolves uniformly about P, the generating point M moves along the line PB in such a manner that the radius vectors shall be inversely proportional to the corresponding arcs, it will describe the *hyperbolic spiral.*

The equation for this spiral is

$$r=\frac{2\pi}{t}, \text{ or } r=2\pi t^{-1}.$$

1. To find the subtangent.

Combining, as above, the equation to the curve with the general equation for the subtangent to polar curves, we obtain

$$PT=-2\pi.$$

Thus in the hyperbolic spiral the *subtangent is constant.*

2. To find the area.

Combining the equation to the curve with the general equation for the area of polar curves, and taking the integral between the limits $t=a$, $t=b$, we obtain

$$A=2\pi^2\left(\frac{1}{a}-\frac{1}{b}\right).$$

3. To find the asymptote.

For $t = 0$, we have $r = \infty$ and $PT = -2\pi$. The spiral has, therefore, an asymptote which will be parallel to the axis from which t is reckoned and at the distance 2π from it.

In general, if a particular value for t renders r infinite, while for the same value the subtangent is either finite or equal to 0, the spiral will have an asymptote drawn through the extremity of the subtangent and parallel to r.

Spirals. Equation $r = at^n$.

130. The spiral of Archimedes and the Hyperbolic spiral are only particular cases of the curves represented generally by the equation $r = at^n$, a being constant and n having all possible values.

1°. Let n be positive. Then if $t = 0$, we shall have $r = 0$, and the spiral passes through the pole.

2°. Let n be negative; the equation then becomes $r = at^{-n}$. Here if $t = 0$, $r = \infty$. In this class of spirals, therefore, the generating point must be considered as commencing its motion at an infinite distance from the pole, toward which it constantly approaches as the radius vector revolves, and which it can reach only after an infinite number of revolutions.

In the general equation $r = at^n$, let $n = 1$, we then have $r = at$, the spiral of Archimedes. In this spiral let $r' =$ the subtangent, then $r' = at^2$, which is the equation of a spiral comprehended in the same general equation. Thus in the spiral of Archimedes the locus of the extremity of the subtangent is a spiral in which t is reckoned from a line 90° distant from the original axis, since r' makes a right angle with r. In like manner the locus of the extremity of the subtangent to the spiral $r' = at^2$ is a spiral, and so on; t in each case being measured from a line 90° distant from that in the preceding curve.

In the general equation $r = at^n$, let $n = -1$, we then have the hyperbolic spiral. If $n = -\frac{1}{2}$, the spiral is called the *Lituus*,

This spiral will have the line from which t is reckoned as an asymptote.

Logarithmic Spiral.

131. If, while the line PD moves uniformly about P, the generating point moves along PA in such a manner that the radius vector increases in a geometric, while the angle increases in an arithmetic ratio, it will describe the *logarithmic spiral.*

From the definition we shall have for its equation

$$t = \log r, \quad \text{or } a^t = r,$$

where r represents the radius vector and t the measuring arc.

1. Subtangent to the curve. For this we obtain

$$\text{PT} = Mr; \text{ whence } \frac{\text{PT}}{r} = M.$$

Thus, *the tangents, at every point, make the same angles with the radius vectors drawn to the point of contact.*

2. Area of the curve. For this we find

$$\text{A} = \tfrac{1}{4}Mr^2 + C,$$

in which the constant will be 0, if the area is reckoned from the pole. If $M = 1$, we shall have $\text{A} = \frac{1}{4}r^2$ for the area of the Naperian logarithmic spiral.

3. Length of the arc. For this we have

$$z = (M^2 + 1)^{\frac{1}{2}} r + C,$$

in which C is 0, if we reckon the arc from the pole.

If $M = 1$, $z = r\sqrt{2}$. In the Naperian logarithmic spiral the length of an arc, therefore, reckoned from the pole, *is equal to the diagonal of the square described upon the radius vector.*

Thus the length of the arc is finite, although the number of revolutions of the radius vector between the pole and its other extremity is infinite.

The logarithmic spiral was invented by Descartes in his investigation of the nature of the line, on which a body acted upon by gravity would be uniformly accelerated, the force of gravity being always directed to the centre of the earth. The line sought, he found to be a spiral having its pole at the centre of the earth.

The full discussion of this curve was, however, reserved for James Bernoulli, who first applied the new Calculus to Geometrical investigations. This spiral was one of the first curves examined with the new instrument. Among the properties immediately discovered was the remarkable one, that the spiral in a great variety of circumstances reproduces itself. Thus it is its own evolute and involute; and if another equal spiral be rolled upon it, the pole of the rolling spiral will describe an equal spiral. On this account Bernoulli called it *spira mirabilis*, and saw in it a type of constancy amid changes, and an emblem of the resurrection.

The most important of the Transcendental curves is the Cycloid, to which we shall attend after providing the means for determining some additional integrals.

SECTION XII.

PROCESSES OF INTEGRATION.

132. The Differential Calculus, considered in itself, is far in advance of the Integral. For there is no quantity the differential of which cannot be found, while there is a large class of differential expressions the integrals of which cannot be obtained; either because they are not the result of an exact differentiation, or because the process for their integration has not yet been discovered.

The problem of integration may be divided into two parts.

1°. To find the values of the *elementary* integrals, or such as cannot be reduced to forms more simple.

2°. To cause a proposed integral to undergo such transformations, that its integration may be made to depend upon some one or more of the elementary forms.

We have already found the elementary integrals both algebraic and transcendental. We now proceed to the second part of the problem.

I. *Change of form by some common algebraic process.*

133. When the proposed expression is not in the form required for integration, it may often, as we have already seen, be reduced to the proper form by some simple algebraic process.

1. Performing operations indicated.

Ex. 1. To find the integral of $(1+x^2)^2dx$.

Performing the operations, the several terms in the result are each integrable, and we have

$$\int(1+x^2)^2dx = x + \tfrac{2}{3}x^3 + \tfrac{1}{5}x^5. \quad \text{Ans.}$$

Ex. 2. Find the integral of $\dfrac{x^3dx}{a-x}$.

Performing the division in relation to x, we have

$$\frac{x^3dx}{a-x} = -x^2dx - axdx - a^2dx + \frac{a^3dx}{a-x};$$

and we shall have for the integral

$$-\tfrac{1}{3}x^3 - \tfrac{1}{2}ax^2 - a^2x - a^3\log(a-x). \quad \text{Ans.}$$

Ex. 3. Find the integral of $\dfrac{x^2dx}{a+x}$.

Ans. $\tfrac{1}{2}x^2 - ax + a^2\log(a+x)$.

2. Dividing an expression into two others of which it is the sum or difference.

Ex. 1. $$\int\frac{(a+bx)dx}{a^2+x^2} = \int\frac{adx}{a^2+x^2} + b\int\frac{xdx}{(a^2+x^2)}$$
$$= \tan^{-1}\frac{x}{a} + \tfrac{1}{2}b\log(a^2+x^2).$$

Ex. 2. $$\int\frac{(1+x)dx}{(a^2-x^2)^{\frac{1}{2}}} = \sin^{-1}\frac{x}{a} - (a^2-x^2)^{\frac{1}{2}}.$$

3. Adding and subtracting the same quantity.

Ex. 1. $\int x(a+x)^{\frac{1}{2}}dx = \int (x+a-a)(a+x)^{\frac{1}{2}}dx$

$$= \int (x+a)^{\frac{3}{2}}dx - a\int (a+x)^{\frac{1}{2}}dx;$$

each of which are known forms.

Ex. 2. $\displaystyle\int \frac{xdx}{(2ax-x^2)^{\frac{1}{2}}} = -\int \frac{(a-x)\,dx}{(2ax-x^2)^{\frac{1}{2}}} + \int \frac{adx}{(2ax-x^2)^{\frac{1}{2}}},$

which are known forms.

4. Multiplying numerator and denominator by the same quantity.

Ex. 1. Find the integral of $\dfrac{(x^2-a^2)^{\frac{1}{2}}dx}{x}$.

Multiplying numerator and denominator by $(x^2-a^2)^{\frac{1}{2}}$, we have

$$\int \frac{(x^2-a^2)^{\frac{1}{2}}dx}{x} = \int \frac{(x^2-a^2)\,dx}{x\,(x^2-a^2)^{\frac{1}{2}}};$$

whence, splitting the fraction into two others, as in the second of the preceding cases, the results reduce to known forms, and we have

$$(x^2-a^2)^{\frac{1}{2}} - a\sec^{-1}\frac{x}{a}. \quad \text{Ans.}$$

Ex. 2. Find the integral of $\dfrac{(a+x)^{\frac{1}{2}}dx}{(a-x)^{\frac{1}{2}}}$.

Ans. $a\sin^{-1}\dfrac{x}{a} - (a^2-x^2)^{\frac{1}{2}}$.

Ex. 3. Find the integral of $\dfrac{(x+a)^{\frac{1}{2}}dx}{x(x-a)^{\frac{1}{2}}}$.

Ans. $\log\left[x+(x^2-a^2)^{\frac{1}{2}}\right] + \sec^{-1}\dfrac{x}{a}$.

II. *Change of form by aid of undetermined coefficients.*

134. Let the expression required to be integrated be a rational fraction in which the numerator is of lower dimensions than the denominator. We may in this case, by the method of undetermined

coefficients, decompose the fraction into a sum of more simple fractions, called *partial fractions;* each of which will be integrable.

We take the two most simple cases.

1. That in which the factors of the denominator are real and unequal.

Ex. 1. Find the integral of $\frac{adx}{x^2 - a^2}$.

The denominator being decomposed into its factors $x - a$, $x + a$, let us put

$$\frac{a}{(x - a)(x + a)} = \frac{A}{x - a} + \frac{B}{x + a},$$

A and B being two constants which it is required to determine. This, it is evident, may be done by the method of undetermined coefficients, Alg. art. 263.

Performing the operations we obtain $A = \frac{1}{2}$, $B = -\frac{1}{2}$; whence by substitution

$$\int \frac{adx}{x^2 - a^2} = \frac{1}{2} \int \frac{dx}{x - a} - \frac{1}{2} \int \frac{dx}{x + a};$$

wherefore

$$\int \frac{adx}{x^2 - a^2} = \frac{1}{2} \log (x - a) - \frac{1}{2} \log (x + a).$$

Ex. 2. To find next the integral of $\frac{(3x - 5)\,dx}{x^2 - 6x + 8}$.

Putting the denominator equal to 0, and resolving the equation thus formed, we obtain $x = 2$, $x = 4$. The factors of the denominator will be, therefore, $x - 2$, $x - 4$, and we put

$$\frac{3x - 5}{(x - 2)(x - 4)} = \frac{A}{x - 2} + \frac{B}{x - 4}.$$

Determining the coefficients we have $A = -\frac{1}{2}$, $B = \frac{7}{2}$.

Whence by substitution and integration, we have for the integral sought

$$\frac{7}{2} \log (x - 4) - \frac{1}{2} \log (x - 2). \quad \text{Ans.}$$

Ex. 3. Find the integral of $\frac{x}{x^2+6x+8}$.

Ans. $2\log(x+4)-\log(x+2)$.

2. We take next the case in which the factors of the denominator of the proposed fraction are real and some of them equal.

Ex. To find the integral of $\frac{x^2dx}{(x-a)^2(x+a)}$.

In order that there may be a sufficient number of equations to determine the coefficients, we must put

$$\frac{x^2}{(x-a)^2(x+a)}=\frac{A}{(x-a)^2}+\frac{A'}{x-a}+\frac{B}{x+a},$$

from which we obtain $A=\frac{1}{2}a$, $A'=\frac{3}{4}$, $B=\frac{1}{4}$; and the integral sought will be

$$-\frac{a}{2(x-a)}+\frac{3}{4}\log(x-a)+\frac{1}{4}\log(x+a). \text{ Ans.}$$

If the fractions are irrational they must be reduced, when it can be done, to a rational form, and the preceding process be then applied. The process of reducing a fraction to a rational form is called *rationalization*. It can be effected in a few particular cases only.

III. *Substitution of a new variable.*

135. Let it be required next to find the integral of

$$x^2(a+x)^{\frac{1}{2}}dx.$$

If the exponent of the parenthesis were an entire number, the whole quantity might, as we have seen, be developed in a limited number of simple terms, each of which would be integrable. Our object is to substitute for the quantity within the parenthesis in such a manner as to render the exponent of the parenthesis entire, and the development, therefore, possible.

Let us assume then, $a+x=z$; we shall have

$$x^2(a+x)^{\frac{1}{2}}dx=z^{\frac{1}{2}}(z-a)^2dz.$$

Developing and integrating, we have for the integral in terms of z,

$$\frac{2}{7}z^{\frac{7}{2}}-\frac{4}{5}az^{\frac{5}{2}}+\frac{2}{3}a^2z^{\frac{3}{2}}=\frac{2}{35}z^{\frac{3}{2}}(5z^2-14az+\frac{35}{3}a^2);$$

or replacing the value of z,

$$\int x^2(a+x)^{\frac{1}{2}}dx=\frac{2}{35}(a+x)^{\frac{3}{2}}(5x^2-4ax^2+\frac{8}{3}a^2).$$

2. The two following cases of substitution are of frequent use, and general formulas for them are easily obtained. We take a few examples of each from which the nature of the method will be easily seen.

Case I. Let it be required to find the integral of

$$x^3(a+bx^2)^{\frac{3}{2}}dx.$$

Assume $a+bx^2=z^2$, then

$$(a+bx^2)^{\frac{3}{2}}=z^3\ (1);\quad x^2=\frac{z^2-a}{b}\ (2);\quad xdx=\frac{zdz}{b}\ (3);$$

whence, multiplying together the equations (1), (2), and (3),

$$x^3(a+bx^2)^{\frac{3}{2}}dx=z^4\frac{z^2-a}{b^2}dz=\frac{z^6dz}{b^2}-\frac{az^4dz}{b^2}.$$

Thus we have obtained for the proposed an equivalent expression, composed of two simple terms; and we have by integration, and restoring the value of z,

$$\int x^3(a+bx^2)^{\frac{3}{2}}dx=\frac{(a+bx^2)^{\frac{7}{2}}}{7b^2}-\frac{a(a+bx^2)^{\frac{5}{2}}}{5b^2}.$$

By a similar substitution we may always integrate a binomial differential, *when the exponent of the variable without the parenthesis, increased by unity, is exactly divisible by the exponent of the variable within.*

Ex. 1. Integrate the expression $du=x^5(a+bx^2)^{\frac{1}{2}}dx$.

$$\text{Ans. } u=\frac{(a+bx^2)^{\frac{7}{2}}}{7b^3}-\frac{2a(a+bx^2)^{\frac{5}{2}}}{5b^3}+\frac{a^2(a+bx^2)^{\frac{3}{2}}}{3b^3}.$$

Ex. 2. Integrate the expression $du = x^3(a - x^2)^{-\frac{1}{2}}dx$.

Ans. $u = -a(a - x^2)^{\frac{1}{2}} + \frac{(a - x^2)^{\frac{3}{2}}}{3}$.

Case II. Let it be proposed next to find the integral of $du = a(1 + x^2)^{-\frac{3}{2}}dx$.

Assume $1 + x^2 = v^2x^2$; then $(1 + x^2)^{-\frac{3}{2}} = v^{-3}x^{-3}$ (1); $x^2 = \frac{1}{v^2 - 1}$, $dx = \frac{-vdv}{x(v^2 - 1)^2}$ (2); and $1 = x^4(v^2 - 1)^2$ (3).

Multiplying together equations (1), (2), and (3), we have

$$du = a(1 + x^2)^{-\frac{3}{2}}dx = -\frac{adv}{v^2};$$

in which the integration of the proposed is made to depend upon a quantity consisting of one simple term only.

Performing the integration and replacing the value of z, we obtain for the integral sought

$$\frac{ax}{(1 + x^2)^{\frac{1}{2}}}. \text{ Ans.}$$

By a similar substitution we may always integrate a binomial differential, *when the exponent of the variable without the parenthesis, augmented by unity, and divided by the exponent of the variable within the parenthesis, plus the exponent of the parenthesis, is a whole number.*

Ex. Integrate the expression $x^{-4}(1 - x^2)^{-\frac{1}{2}}dx$.

Ans. $\frac{1 + 2x^2}{3x^3}(1 - x^2)^{\frac{1}{2}}$.

IV. *Integration by parts.*

136. We have for the differential of a product uv,

$$d(uv) = udv + vdu,$$

whence

$$\int udv = uv - \int vdu$$

By means of this formula we may often decompose a required integral into two parts, one already integrated and the other easy to be integrated.

Ex. 1. To find the integral of $x^3dx\,(a^2-x^2)^{\frac{1}{2}}$.

Put $x^2=u$, and $xdx\,(a^2-x^2)^{\frac{1}{2}}=dv$; then by the formula

$$\int x^2(a^2-x^2)^{\frac{1}{2}}xdx=-\tfrac{1}{3}x^2(a^2-x^2)^{\frac{3}{2}}+\tfrac{2}{3}\int(a^2-x^2)^{\frac{3}{2}}xdx;$$

whence $\int x^3dx\,(a-x^2)^{\frac{1}{2}}=-\tfrac{1}{3}x^2(a^2-x^2)^{\frac{3}{2}}-\tfrac{2}{15}(a^2-x^2)^{\frac{5}{2}}$.

Ex. 2. To find the integral of $(a^2-x^2)^{\frac{1}{2}}dx$.

Put $(a^2-x^2)^{\frac{1}{2}}=u$, and $dx=dv$.

Then $\int(a^2-x^2)^{\frac{1}{2}}dx=x\,(a^2-x^2)^{\frac{1}{2}}+\int\frac{x^2dx}{(a^2-x^2)^{\frac{1}{2}}}$. (1)

Multiplying and dividing the proposed by $(a^2-x^2)^{\frac{1}{2}}$, we obtain the identical equation

$$\int(a^2-x^2)^{\frac{1}{2}}dx=\int\frac{a^2dx}{(a^2-x^2)^{\frac{1}{2}}}-\int\frac{x^2dx}{(a^2-x^2)^{\frac{1}{2}}};\quad(2)$$

adding (1) and (2) we obtain

$$2\int(a^2-x^2)^{\frac{1}{2}}dx=x\,(a^2-x^2)^{\frac{1}{2}}+\int\frac{a^2dx}{(a^2-x^2)^{\frac{1}{2}}},$$

which reduces the integration of the proposed to that of a known form, and we have finally

$$\int(a^2-x^2)^{\frac{1}{2}}dx=\tfrac{1}{2}x\,(a^2-x^2)^{\frac{1}{2}}+\tfrac{1}{2}a^2\sin^{-1}\frac{x}{a}.$$

Ex. 3. To find the integral of $\frac{x^2dx}{(a^2-x^2)^{\frac{1}{2}}}$.

Ans. $-\tfrac{1}{2}x\,(a^2-x^2)^{\frac{1}{2}}+\tfrac{1}{2}a^2\sin^{-1}\frac{x}{a}$.

V. *Method of Reduction.*

137. After the first division of a proposed integral into parts, the part to be integrated may again be divided into parts; and so on, until, in the cases to which the formula is applicable, a final integral sufficiently simple is obtained. The process is called integration by *continuation* or *reduction*.

The method of reduction which we have now to explain is different from this. It consists in transforming, by aid of certain assumptions, a proposed integral into one more simple, and this again, by a repetition of the process, into one yet more simple, until it is made to depend upon the most simple of its class.

The first form, to which the method applies, and the one of most frequent occurrence, is the binomial form $\int x^{m-1}dx(a+bx^n)^p$. The method depends upon the principle that we may always make an integral of this form depend upon another of the same form in which one of the indices is altered; viz. that of x without the parenthesis by the addition or subtraction of n, and that of the parenthesis by the addition or subtraction of 1.

Without preparing the general formulas, we shall solve a few particular cases, which will be sufficient to show the general spirit of the method, and furnish the means of obtaining some additional integrals we shall have occasion to use.

Ex. 1. Let it be required to find the integral of

$$x^3dx\,(a^2-x^2)^{-\frac{1}{2}}.$$

In this case the exponent of x without the parenthesis must be diminished. We assume, therefore, $x^2(a^2-x^2)^{\frac{1}{2}}=u$, from which we obtain by differentiation

$$2xdx(a^2-x^2)^{\frac{1}{2}}-x^3dx\,(a^2-x^2)^{-\frac{1}{2}}=du.$$

Multiplying and dividing the first term of the left hand member by $(a^2-x^2)^{\frac{1}{2}}$, transposing and reducing we obtain

$$3x^3dx(a^2-x^2)^{-\frac{1}{2}}=2a^2xdx(a^2-x^2)^{-\frac{1}{2}}-du.$$

The integration, therefore, is now made to depend upon that of $xdx(a^2-x^2)^{-\frac{1}{2}}$, of the same form with the proposed but more simple. And this being a known form, the integration is thus completely effected. Integrating and reducing we obtain

$$\int\frac{x^3dx}{(a^2-x^2)^{\frac{1}{2}}}=-\frac{1}{3}\,(2a^2+x^2)\,(a^2-x^2)^{\frac{1}{2}}.$$

Ex. 2. Find the integral of $x^2dx\,(1-x^2)^{-\frac{1}{2}}$.

Assume $x\,(1-x^2)^{\frac{1}{2}}=u$. Ans. $\frac{1}{2}\sin^{-1}x-\frac{1}{2}x(1-x^2)^{\frac{1}{2}}$.

Ex. 3. Find the integral of $xdx\,(1+x)^{\frac{1}{2}}$.

Assume $x(1+x)^{\frac{3}{2}}=u$. Ans. $\frac{2}{3}\,(1+x)^{\frac{3}{2}}\cdot\frac{3x-2}{5}$.

Ex. 4. Find the integral of $dx\,(a^2+x^2)^{-3}$.

In this case, the exponent of the parenthesis must be increased. We assume, therefore, $x\,(a^2+x^2)^{-2}=u$.

Differentiating we obtain

$$du=dx\,(a^2+x^2)^{-2}-4x^2dx\,(a^2+x^2)^{-3};$$

adding and subtracting $4a^2dx\,(a^2+x^2)^{-3}$,

$$du=dx\,(a^2+x^2)^{-2}-4dx\,(a^2+x^2)\,(a^2+x^2)^{-3}+4a^2dx\,(a^2+x^2)^{-3};$$

or
$$du=-3dx\,(a^2+x^2)^{-2}+4a^2dx\,(a^2+x^2)^{-3};$$

whence, transposing and integrating

$$\int\frac{dx}{(a^2+x^2)^3}=\frac{x}{4a^2(a^2+x^2)^2}+\frac{3}{4a^2}\int\frac{dx}{(a^2+x^2)^2};$$

and the proposed integral now depends upon $\int dx\,(a^2+x^2)^{-2}$, a more simple integral of the same class. Repeating the process upon this last, we obtain

$$\int\frac{dx}{(a^2+x^2)^2}=\frac{1}{2a^2}\cdot\frac{x}{a^2+x^2}+\frac{1}{2a^3}\cdot\tan^{-1}\frac{x}{a},$$

by means of which the integration of the proposed is completely effected.

Ex. 5. Find the integral of $x^{-3}dx\,(a^2+x^2)^{-\frac{1}{2}}$.

Here the exponent of x without the parenthesis must be increased. We assume, therefore, $x^{-2}(a^2+x^2)^{\frac{1}{2}}=u$, and obtain

$$\int\frac{dx}{x^3(a^2+x^2)^{\frac{1}{2}}}=-\frac{1}{2a^2}\,\frac{(a^2+x^2)^{\frac{1}{2}}}{x^2}-\frac{1}{2a^2}\int\frac{dx}{x\,(a^2+x^2)^{\frac{1}{2}}};$$

and thus arrive at a known form.

Ex. 6. Find the integral of $dx(a^2 - x^2)^{\frac{3}{2}}$.

Here the exponent of the parenthesis must be diminished, and we assume $x(a^2 - x^2)^{\frac{3}{2}} = u$.

Differentiating we obtain

$$du = dx(a^2 - x^2)^{\frac{3}{2}} - 3x^2dx(a^2 - x^2)^{\frac{1}{2}},$$

$$= dx(a^2 - x^2)^{\frac{1}{2}}[(a^2 - x^2) - 3x^2],$$

$$= dx(a^2 - x^2)^{\frac{1}{2}}[(a^2 - x^2) + 3(a^2 - x^2) - 3a^2];$$

from which we derive

$$\int dx(a^2 - x^2)^{\frac{3}{2}} = \frac{1}{4}x(a^2 - x^2)^{\frac{3}{2}} + \frac{3}{4}a^2\int dx(a^2 - x^2)^{\frac{1}{2}};$$

and the integral now depends upon a quantity already integrated, art. 136.

Ex. 7. Find the integral of $dx(2ax - x^2)^{\frac{1}{2}}$. This is comprehended in the general form above, since the factor x within the parenthesis may be placed without.

Assume $x(2ax - x^2)^{\frac{1}{2}} = u$. Differentiating and reducing we obtain

$$dx(2ax - x^2)^{\frac{1}{2}} = -(ax - x^2)(2ax - x^2)^{-\frac{1}{2}}dx + du;$$

$$= -dx(2ax - x^2)^{\frac{1}{2}} + axdx(2ax - x^2)^{-\frac{1}{2}} + du;$$

$$2dx(2ax - x^2)^{\frac{1}{2}} = -a(a - x)dx(2ax - x^2)^{-\frac{1}{2}} + a^2dx(2ax - x^2)^{-\frac{1}{2}} + du;$$

$$\int dx(2ax - x^2)^{\frac{1}{2}} = \frac{x - a}{2}(2ax - x^2)^{\frac{1}{2}} + \frac{a^2}{2}\text{ver-sin}^{-1}\frac{x}{a}.$$

Ex. 8. Find the integral of $x^2dx(2ax - x^2)^{-\frac{1}{2}}$.

Assume $x(2ax - x^2)^{\frac{1}{2}} = u$.

Ans. $\frac{3}{2}a^2\text{ver-sin}^{-1}\frac{x}{a} - \frac{3}{2}a(2ax - x^2)^{\frac{1}{2}} - \frac{1}{2}x(2ax - x^2)^{\frac{1}{2}}$.

Ex. 9. Find the integral of $x^3dx\,(1+x^2)^{\frac{1}{2}}$.

$$\text{Ans. } \frac{1}{15}(1+x^2)^{\frac{3}{2}}(3x^2-2).$$

VI. *Integration by Infinite Series.*

138. When the proposed differential cannot be integrated by the ordinary methods, it must be converted into an infinite series and each term separately integrated. If the series is converging, the value of the integral, it is evident, may be obtained with such degree of approximation as we please.

The following examples will sufficiently illustrate the method.

Ex. 1. Let it be required to find the integral of $\frac{dx}{a+x}$.

Converting into series we have

$$\frac{1}{a+x}=\frac{1}{a}-\frac{x}{a^2}+\frac{x^2}{a^3}-\frac{x^3}{x^4}+\&c.$$

Multiplying by dx and integrating

$$\int\frac{dx}{a+x}=\frac{x}{a}-\frac{x^2}{2a^2}+\frac{x^3}{3a^3}-\frac{x^4}{4a^4}+\&c.+C.$$

Recollecting that the left hand member $=\log(a+x)$, and determining the constant by the hypothesis $x=0$, we obtain the following series for $\log(a+x)$, viz.

$$\log(a+x)=\log a+\frac{x}{a}-\frac{x^2}{2a^2}+\frac{x^3}{3a^3}-\frac{x^4}{4a^4}+\&c.$$

Ex. 2. Find the integral of $\frac{dx}{1+x^2}$.

$$\text{Ans. } x-\frac{x^3}{3}+\frac{x^5}{5}-\frac{x^7}{7}+\&c.$$

Circular area forms.

139. By means of series we may find, to any degree of approximation we please, the values of the following integrals.

$$(1)\ \int dx\,(a^2-x^2)^{\frac{1}{2}}; \qquad (2)\ \int dx\,(2ax-x^2)^{\frac{1}{2}};$$

the values of which we have already determined by the processes of arts. 136, 137.

These integrals express, it is easy to see, the area of a circular segment to the radius a; the first when the origin of abscissas is taken at the centre of the circle, and the second when it is taken upon the circumference.

The values of the circular area forms to the radius 1, have been calculated and registered in tables. The above forms may, therefore, be added to the elementary forms, art. 120, since an integral may be regarded as fully determined, when it is made to depend either upon the one or the other of them.

Tables of elliptic and hyperbolic forms, as well as of some others, have also been computed, by means of which the process of integration is, in like manner, facilitated.

With the increased means of integration now obtained we resume the applications of the Calculus, commencing as proposed with the Cycloid.

SECTION XIII.

TRANSCENDENTAL CURVES CONTINUED. THE CYCLOID.

140. Next to the conic sections the cycloid has received the greatest attention from mathematicians, and the result has been the discovery of numerous properties of great importance both in geometry and dynamics. The investigation of these properties is well adapted to illustrate the power of the Calculus. To the more important of them we now proceed, observing in general the order in which, by far inferior means, they were first discovered.

141. The cycloid was first conceived by Galileo, who, from its graceful form, thought it suitable for the arches of a bridge. It is defined as follows,

If a circle GMR be rolled along a straight line AL, any point M of the circumference of this circle will describe a curve, which is called the *cycloid*. The circle GMR is called the *generating circle*, and the point M the *generating point*.

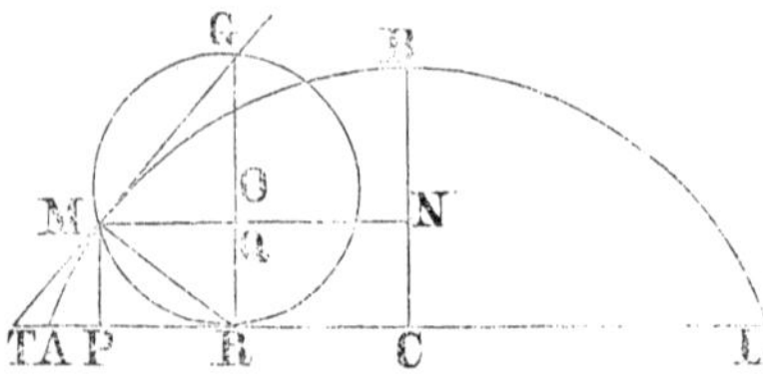

If the point M is at A, on the line AL, at the commencement of the motion, at the end of one revolution of the generating circle, it will again be found on the line AL, at some point L. And since all the points of the generating circle will thus have come successively in contact with the line AL, this line will be equal to the circumference of the generating circle. It is called the *base of the cycloid*. The line BC drawn perpendicular to the base, at the middle point C, is equal to the diameter of the generating circle, and is called the *axis of the cycloid*.

If after one revolution the generating circle continues to roll along the line AL, another equal curve will be described. We shall consider only the curve described in the first revolution.

142. To find an equation for the curve, let A be assumed as the origin of the co-ordinates. Let us suppose that the generating point M has described the arc AM, and that R is the point of contact of the generating circle with the base AL. From R draw the diameter RG perpendicular to AL; and from M draw MP perpendicular and MQ parallel to AL. Then AR will be equal, it is evident, to the arc RM, and PM will be equal to RQ which is the ver-sine of the arc RM.

Let $r =$ OR, the radius of the generating circle, AP $= x$, PM $=$ RQ $= y$. We shall then have

$$x = \text{AR} - \text{PR} = \text{arc RM} - \text{PR};$$

but, from the nature of the circle, PR $=$ MQ $= (2ry - y^2)^{\frac{1}{2}}$; also the arc RM = arc whose versed sine is RQ or y; substituting these values, we shall have for the equation to the cycloid

$$x = \text{ver-sin}^{-1} y - (2ry - y^2)^{\frac{1}{2}},$$

in which the ver-sin is to the radius r.

Differentiating both sides of this equation and reducing we obtain

$$dx = \frac{y dy}{(2ry - y^2)^{\frac{1}{2}}}, \quad (1)$$

the differential equation of the cycloid when the base is taken for the axis of x.

2. Taking next the axis BC for the axis of x and the vertex B for the origin, let BN $= x$, NM $= y$. Then

$$y = \text{MO} + \text{RC} = (2rx - x^2)^{\frac{1}{2}} + \text{RC}.$$

But RC = AC — AR $= \pi r$ — arc MR; whence

$$y = (2rx - x^2)^{\frac{1}{2}} + \pi r - \text{ver-sin}^{-1} (2r - x);$$

from which by differentiation and reduction we obtain

$$dy = \frac{(2rx - x^2)^{\frac{1}{2}}}{x} dx, \quad (2)$$

the differential equation of the cycloid, when the axis BC of the curve is taken for the axis of x, and the origin is at B.

Area of the Cycloid.

143. Galileo suspected the area of the cycloid to be equal to three times the area of the generating circle. This he sought to verify by a method purely mechanical. Failing to accomplish the object, he gave up the problem as hopeless. About the year 1628, Mersenne suggested the problem to Roberval, one of the most distinguished geometers of the day, who failed to solve it. After a period of six years, devoted to the study of the ancient geometry, the attention of Roberval was again directed to the problem, when he succeeded in effecting a solution. Let us now apply the Calculus to it.

Resuming the general element of a curve surface $dA = ydx$, and integrating this expression by parts, we obtain

$$A = xy - \int xdy;$$

substituting for dy its value from equation (2) of the cycloid, we obtain for the area

$$A = xy - \int dx\,(2rx - x^2)^{\frac{1}{2}} + C,$$

in which $\int dx\,(2rx - x^2)^{\frac{1}{2}}$ expresses the circular area to the radius r and abscissa x.

If we take the integral between $x = 0$, $x = 2r$, recollecting that when $x = 2r$, $y = AC = \pi r$, we obtain

$A = 2\pi r^2$ — semicircle BGC

$= 2\pi r^2 - \frac{1}{2}\pi r^2 = \frac{3}{2}\pi r^2$;

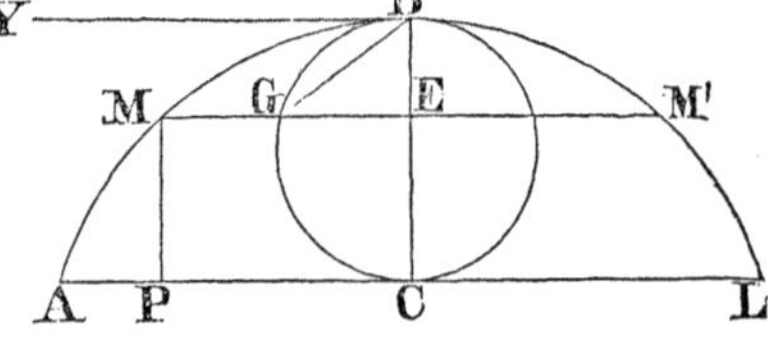

whence, doubling to obtain the whole area, we have *the area of the cycloid equal to three times the area of the generating circle.*

Tangent and Normal.

144. The success of Roberval in discovering the quadrature of the cycloid was communicated by Mersenne to Descartes, then in possession of the analytic or new geometry of which he was the founder. Descartes, disposed to speak slightingly of the discovery of Roberval, proposed as a challenge the problem of the tangent to the curve. Roberval failed to solve the problem, but Fermat, a geometer little inferior to Descartes, succeeded in obtaining a solution.

We now apply the Calculus to this problem.

1. Resuming the general expression for the subtangent, art. 30, and substituting from equation (1) of the cycloid, we obtain

$$\text{Subtan. PT} = \frac{y^2}{(2ry - y^2)^{\frac{1}{2}}};$$

the base being the axis of x.

2. By a similar substitution we obtain for the subnormal

$$\text{Subnormal PR} = (2ry - y^2)^{\frac{1}{2}}.$$

But this (fig. art. 142) is equal to MQ. Thus *the normal MR and the diameter RG of the generating circle intersect the base at the same point.*

This property furnishes a ready method for drawing a tangent to any point M of the curve.

3. To determine next where the tangent is perpendicular to or parallel to the base of the curve, from the equation to the curve we deduce

$$\frac{dy}{dx} = \frac{(2ry - y^2)^{\frac{1}{2}}}{y}.$$

Here if we put $y = 0$, $\frac{dy}{dx} = \infty$.

Thus the tangent is perpendicular to the base at the origin.

Again, if we make $y = 2r$, we have $\frac{dy}{dx} = 0$.

The tangent is therefore parallel to the base when y is equal to $2r$, or at the extremity of the axis.

4. Since for $y = 2r$, $\frac{dy}{dx} = 0$, *the maximum value of* y *will be the diameter of the generating circle.*

Volume of the solids of Revolution.

145. The next problem in regard to the cycloid which obtained solution, relates to the solids formed by the revolution of the curve about its base and axis. This was accomplished about the year 1644, and the honor of being first to obtain the solution is incontestibly due to Roberval.

Let it be proposed next to find the volume of the solid produced by the revolution of the cycloid about its base.

Resuming the general element for the solids of revolution,

$dV = \pi y^2 dx$, and substituting from the equation of the curve we obtain

$$V = \int \frac{\pi y^3 dy}{(2ry - y^2)^{\frac{1}{2}}}.$$

Employing, to obtain the integral, the process of art. 137, we assume $y^2(2ry - y^2)^{\frac{1}{2}} = u$, or placing the y^2 within the parenthesis, $(2ry^5 - y^6)^{\frac{1}{2}} = u$.

Differentiating, transposing, reducing and integrating, we obtain

$$\int \frac{\pi y^3 dy}{(2ry - y^2)^{\frac{1}{2}}} = \frac{5}{3} \pi r \int \frac{y^2 dy}{(2ry - y^2)^{\frac{1}{2}}} - \frac{1}{3} \pi y^2 (2ry - y^2)^{\frac{1}{2}}.$$

$$\text{Also, } \int \frac{y^2 dy}{(2ry - y^2)^{\frac{1}{2}}} = \frac{3}{2} r^2 \text{ ver-sin}^{-1} \frac{y}{r} - \frac{3}{2} r (2ry - y^2)^{\frac{1}{2}} \ldots$$
$$- \frac{1}{2} y (2ry - y^2)^{\frac{1}{2}}.$$

If we now take the integral between $y = 0$, $y = 2r$, and double the result, we have for the whole solid

$$2V = 5\pi^2 r^3.$$

But the base of the circumscribed cylinder is equal, it is evident, to $\pi(2r)^2$, and its solidity to $8\pi^2 r^3$. The solid, therefore, generated by the revolution of the cycloid about its base *is equal to five-eighths of the circumscribed cylinder.*

Segments of the Cycloid.

146. Twelve years after the last of the preceding discoveries the attention of mathematicians was again directed to the cycloid by the problems proposed by Pascal. Thus far the area of the curve and the volume of its solids had been found for the whole curve only. Pascal proposed the more general problem, viz. to determine any segment of the area, and of the solids of revolution; and to the solution challenged the mathematicians of the day.

The preceding solutions embrace this problem, so far as it relates to the area of the curve, and the solid produced by its revolution about the base.

Thus if we wish to find the area of the segment MBM′ (fig. art. 143), resuming the expression for the area

$$A = xy - \int dx\,(2rx - x^2)^{\frac{1}{2}}, \quad (1)$$

we have only to take the integral between $x = 0$, $x = \text{BE}$. Twice the value of this integral will be the area sought.

1. As a particular case let $x = \text{BE} = \frac{1}{2}r$. Then

$$y = \text{GE} + \text{MG}, \text{ arc BG} = 60^\circ = \tfrac{1}{3}\pi r.$$

But $\text{GE} = \sqrt{(\text{BE} \times \text{EC})} = \sqrt{(\frac{3}{4}r^2)} = \frac{1}{2}r\sqrt{3}$; and, by the property of the curve, $\text{MG} =$ the arc $\text{BG} = \frac{1}{3}\pi r$; whence $y = \frac{1}{2}r\sqrt{3} + \frac{1}{3}\pi r$; wherefore

$$xy = \tfrac{1}{2}r(\tfrac{1}{2}r\sqrt{3} + \tfrac{1}{3}\pi r) = \tfrac{1}{4}r^2\sqrt{3} + \tfrac{1}{6}\pi r^2;$$

also, art. 137, Ex. 7, $\int dx(2rx - x^2)^{\frac{1}{2}} = -\frac{1}{8}r^2\sqrt{3} + \frac{1}{6}\pi r^2$;

whence by substitution in (1) $2A = \frac{3}{4}r^2\sqrt{3}$.

But $\text{GE} \times \text{EC} = \frac{3}{4}r^2\sqrt{3}$. Thus the area of the segment is equal, in this case, to the rectangle $\text{GE} \times \text{EC}$.

2. If $x = r$ we find, in like manner, the area of the segment equal r^2.

The first of these properties was discovered by Huygens, and also by Wren in the researches occasioned by the challenge of Pascal. The most important result of these investigations was, however, the discovery by Wren of the rectification of the cycloid, and the quadrature of the surface of its solids of revolution.

Rectification of the Cycloid.

147. Resuming the expression for the element of a curve line

$$dz = (dx^2 + dy^2)^{\frac{1}{2}},$$

substituting from equation (2) of the cycloid and reducing, we obtain

$$dz = \left(\frac{2r}{x}\right)^{\frac{1}{2}} dx;$$

whence by integration

$$z = 2(2rx)^{\frac{1}{2}} + C.$$

Taking the integral between $x = 0$, $x = \text{BE}$ (fig. art. 143), we have

$$z = 2\,(2r\text{BE})^{\frac{1}{2}} = 2\,\text{BG},$$

since BG is a mean proportional between BC and BE.

Thus, *the arc of the cycloid estimated from the vertex of its axis is equal to twice the corresponding chord of the generating circle.*

Taking the integral between $x = 0$, $x = 2r$, we have $z = 2(2r)$ or twice the diameter of the generating circle. Thus, *the whole length of the cycloid is four times the diameter of the generating circle.*

Surface of the Solids of Revolution.

148. 1. To find the surface of the solid produced by the revolution of the cycloid about its base.

Resuming the general element of the surfaces of revolution

$$ds = 2\pi y\,(dx^2 + dy^2)^{\frac{1}{2}},$$

and substituting from equation (1) of the cycloid, and reducing, we obtain

$$ds = 2\pi\,(2r)^{\frac{1}{2}}\,\frac{ydy}{(2r - y)^{\frac{1}{2}}}.$$

Integrating by the method of art. 137, we have

$$\text{S} = 2\pi(2r)^{\frac{1}{2}}\,[-\tfrac{4}{3}\,(2r - y)^{\frac{3}{2}} - 2y(2r - y)^{\frac{1}{2}}] + C.$$

Taking the integral from $y = 0$ to $y = 2r$, and extending the integral over the whole surface, we have

$$2\,\text{S} = \frac{64}{3}\,\pi r^2.$$

When the cycloid is revolved about its base, the whole surface, therefore, *is equal to sixty-four thirds of the generating circle.*

2. To find the surface of the solid when the curve is revolved about its axis.

Substituting in the general element from equation (2) of the cycloid, we have

$$ds = 2\pi\,(2r)^{\frac{1}{2}}\,yx^{-\frac{1}{2}}\,dx.$$

Integrating by parts, and substituting for dy its value

$$S = 4\pi y(2rx)^{\frac{1}{2}} - 4\pi(2r)^{\frac{1}{2}}\int dx\,(2r-x)^{\frac{1}{2}}.$$

Integrating from $x=0$ to $x=2r$, and recollecting that when $x=2r$, $y=\pi r$, we have for the surface

$$S = 8\pi r^2\left(\pi - \frac{4}{3}\right).$$

Radius of Curvature. Evolute.

149. If we differentiate the equation to the cycloid

$$dx = \frac{ydy}{(2ry-y^2)^{\frac{1}{2}}},$$

regarding dx as constant, we obtain

$$0 = (yd^2y + dy^2)\,(2ry - y^2)^{\frac{1}{2}} - \frac{ydy\,(rdy - ydy)}{(2ry-y^2)^{\frac{1}{2}}},$$

or reducing and dividing by y,

$$0 = (2ry - y^2)\,d^2y + rdy^2;$$

whence we obtain $d^2y = -\dfrac{rdy^2}{2ry - y^2} = -\dfrac{rdx^2}{y^2}$.

Substituting next the values of dy and d^2y in the general expression for the radius of curvature,

$$R = \frac{(dx^2 + dy^2)^{\frac{3}{2}}}{dxd^2y},$$

art. 102, and reducing, we obtain

$$R = 2(2ry)^{\frac{1}{2}}.$$

But if we determine the value of the normal MR (fig. art. 141) we shall find $MR = (2ry)^{\frac{1}{2}}$. Thus *the radius of curvature corresponding to any point of the cycloid, is equal to double the normal.*

Substituting next in the formulas for the co-ordinates of the centre of the osculatory circle, equations (7) and (8) art. 102, the values of dy, d^2y already found, we obtain

$$y = -b, \ x = a - 2(2ry - y^2)^{\frac{1}{2}}.$$

Substituting these values in the transcendental equation of the cycloid, $x = \text{ver-sin}^{-1} y - (2ry - y^2)^{\frac{1}{2}}$, we obtain

$$a = \text{ver-sin}^{-1}(-b) + (-2rb - b^2)^{\frac{1}{2}},$$

the equation of the evolute referred to the primitive origin and the primitive axes. It is also the equation of a cycloid situated below the axis of x, the generating circle of which is equal to that of the given one.

If the origin be transferred to A′, A′C being taken equal to CB or $2r$, the vertex B will be transferred to A, the extremity of the base, and the arc A′A will be identical with the arc AB. Thus *the evolute of the cycloid is an equal cycloid.*

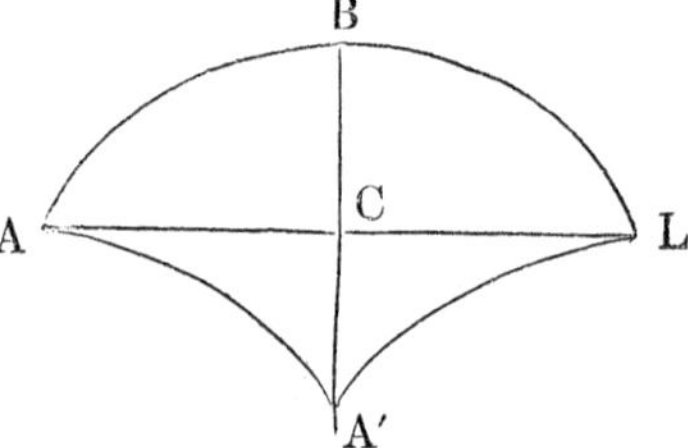

These properties were first discovered by Huygens.

Dynamical properties.

150. In addition to the preceding geometrical properties, the cycloid is found to possess the two following remarkable dynamical properties.

1°. It is the curve along which a body under the action of gravity, will fall in the shortest time from one point to another not in the same vertical.

2°. If the cycloid be inverted with its base horizontal, a body under the action of gravity will reach the lowest point in the same time from whatever point of the curve it begins to fall.

We shall resume these problems when we come to the application of the Calculus to Mechanics.

SECTION XIV.

QUADRATURE OF AREAS AND CUBATURE OF SOLIDS CONTINUED.

151. We have now seen the manner in which the Calculus is applied to the various problems which relate to plane curves. We proceed to some further examples of its application to the quadrature of areas bounded by curve lines and the cubature of the solids produced by their revolution.

Ex. 1. *Cissoid of Diocles.*

This curve was invented by Diocles, a Greek geometer who lived about the sixth century of the Christian era. The purpose of its invention was the solution of the problem of finding two mean proportionals. The curve is generated as follows. On the diameter AB of a circle erect two equal ordinates NR, MQ. Join QA; the point P in which the line QA intersects NR will be a point of the curve. In like manner, through A and R draw a line meeting in a point P′ the ordinate MQ produced; P′ will be a point in the curve.

To find an equation to the curve let $AB=2a$, $AN=x$, $PN=y$. Then, by similar triangles,

$$\frac{PN}{AN}=\frac{QM}{AM}, \text{ or } \frac{y}{x}=\frac{(2ax-x^2)^{\frac{1}{2}}}{2a-x};$$

and we shall have for the equation sought

$$y^2=\frac{x^3}{2a-x}.$$

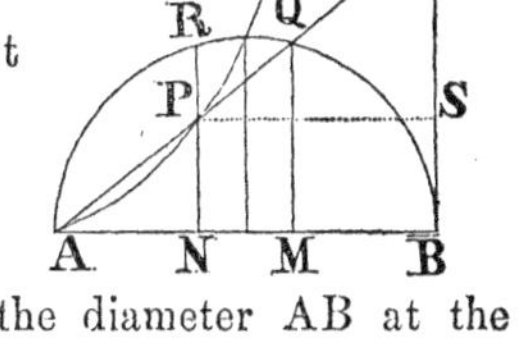

Figure of the curve. 1°. It meets the diameter AB at the point A. 2°. It passes through the point D, the extremity of the radius CD. 3°. The perpendicular BM, at the extremity of the diameter AB is an asymptote to the curve. 4°. It consists of two equal branches, one above and the other below the diameter AB, and which meet in a cusp at the point A. This last property, through the defect of their geometry, was not perceived by the ancient geometers.

1. To find the area.

Substituting in the general element, $dA = ydx$, we have for the area of the curve

$$A = \int \frac{x^{\frac{3}{2}} dx}{(2a - x)^{\frac{1}{2}}}.$$

This integral is easily reduced to Ex. 8, art. 137. Integrating, however, by parts we have

$$A = \int x^{\frac{3}{2}} \frac{dx}{(2a - x)^{\frac{1}{2}}} = -2x^{\frac{3}{2}}(2a - x)^{\frac{1}{2}} + 3\int dx(2ax - x^2)^{\frac{1}{2}} + C,$$

or area $\text{ANP} = -2x^{\frac{3}{2}}(2a - x)^{\frac{1}{2}} + 3$ circular area $\text{ANR} + C$.

Taking the integral between $x = 0$, $x = 2a$, the area between the curve and its asymptote will be three times the area of the semicircle ARB. Or, doubling, on account of the two equal branches, the whole area of the curve will be *three times the area of the generating circle.*

1. Volume of the solids produced by the revolution of the curve.

First, if the curve revolve about the axis AB, we have, by substitution in the general element, $\pi y^2 dx$, of the solids of revolution, dividing and integrating

$$V = \int \frac{\pi x^3 dx}{2a - x} = \pi \left(-\frac{x^3}{3} - ax^2 - 4a^2 x + 8a^3 \log \frac{2a}{2a - x} \right);$$

the constant being determined by reckoning the integral from the vertex A.

Second, if the curve revolve about the asymptote, we have

$$dV = \pi(\text{PS})^2 dy = \pi (2a - x)^2 \frac{(3a - x) x^{\frac{1}{2}}}{(2a - x)^{\frac{3}{2}}} dx,$$

$$= \pi (3a - x) dx (2ax - x^2)^{\frac{1}{2}};$$

from which we obtain by integration

$$V = \pi\left[2a\int dx\,(2ax - x^2)^{\frac{1}{2}} + \frac{1}{3}(2ax - x^2)^{\frac{3}{2}}\right] + C.$$

Taking the integral between $x = 0$, $x = 2a$, we have

$$V = \pi\,(2a\,.\,\tfrac{1}{2}\,\pi\,a^2) = \pi^2 a^3,$$

or doubling for both branches $2V = 2\pi^2 a^3$.

Ex. 2. *The Witch of Agnesi.*

This curve was the invention of an Italian lady, Donna Maria Agnesi. It is described as follows. In the ordinate produced of a circle AMB, take a point P such that PN : AB : : MN : AN ; P will be a point of the curve. Putting $AC = a$, $AN = x$, $PN = y$, the equation will be

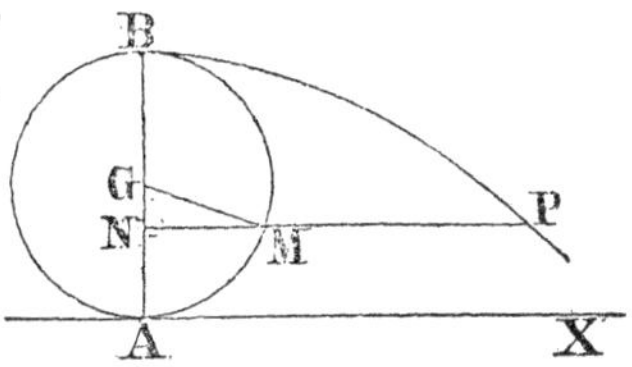

$$xy^2 = 4a^2(2a - x).$$

1. To find the area we have

$$dA = 2ax^{-\frac{1}{2}}dx\,(2a - x)^{\frac{1}{2}};$$

whence by the process, No. 4, art. 133,

$$A = 2a\left[(2ax - x^2)^{\frac{1}{2}} + a\,\text{ver-sin}^{-1}\frac{x}{a}\right] + C.$$

Taking this from $x = 2a$ to $x = 0$, and doubling, we have

$$2A = 4\pi a^2.$$

Hence the whole area between the curve and its asymptote, will be equal to *four times the area of the generating circle.*

2. To find the volume of the solid produced by the revolution of the curve about its asymptote.

Taking the asymptote AK for the axis of x, the equation to the curve will be $xy = 2a(2ay - y^2)^{\frac{1}{2}}$.

But, integrating by parts,

$$V = \pi\int y^2dx = \pi y^2x - 2\pi\int xydy,$$

whence by substitution

$$V = \pi y^2x - 4\pi a\int dy\,(2ay - y^2)^{\frac{1}{2}} + C;$$

and we shall have for the whole volume $4\pi^2a^3$.

Ex. 3. *The companion to the Cycloid.*

This curve derives its origin from the cycloid. If, in the preceding figure, an ordinate NM of a circle be produced until we have NP equal to the arc BM, intercepted between it and the extremity of the diameter, the extremity P of this ordinate will be a point in the curve.

Putting $AC = a$, $PN = x$, $AN = y$, the angle $BCM = \omega$, the curve will be defined by the equations

$$x = a\omega, \qquad (1)$$

$$y = a(1 + \cos \omega). \quad (2)$$

To find the area, we take the differential of the first equation and multiply the result by the second, which gives by integration

$$A = a^2(\omega + \sin \omega) + C.$$

Taking this between $\omega = 0$, $\omega = \pi$, and doubling, we have for the whole area $$2A = 2\pi a^2.$$

Thus the whole area of the curve is equal to *twice the area of the generating circle.*

Ex. 4. *The Lemniscate of Bernoulli.*

The form of this curve is that of the figure 8. It has for its polar equation $$r^2 = a^2 \cos 2\omega.$$

To find the area, we have by substitution in the element of the area of polar curves, $dA = \frac{1}{2}r^2 dt$, art. 126, and integrating,

$$A = \tfrac{1}{2}a^2 \int \cos 2\omega d\omega = \tfrac{1}{4}a^2 \int \cos 2\omega 2d\omega = \tfrac{1}{4}a^2 \sin 2\omega + C.$$

If we take this from $\omega = 0$ to $\omega = \frac{1}{4}\pi$, we have

$$A = \tfrac{1}{4}a^2.$$

This is the fourth part of the area of the whole curve, which is, therefore, equal to a^2.

Ex. 5. To find the whole area of the curve expressed by the equation $$y(a^2 - x^2)^{\frac{1}{2}} = 4x^2.$$

Ans. $4a^2\pi$, or four times the area of the circle whose radius is a.

Ex. 6. Find the area of the curve expressed by the equation $y(a^2 - x^2) = a^3$.

Ans. $\frac{a^2}{2} \log \frac{a + x}{a - x}$.

DOUBLE INTEGRALS.

152. When an area is referred to rectangular co-ordinates, the most general expression for its element is the infinitely small rectangle of which the differentials dx and dy are the sides. Thus we have

$$dA = dydx.$$

To find the area we must integrate between proper limits with respect to each of the variables in succession. This double integration is indicated by writing the sign of integration twice before the quantity to be integrated. Thus we have

$$A = \int\int dydx.$$

One of these integrations may always be performed, so that we shall have either

$$\int ydx + C, \text{ or } \int xdy + C;$$

and these integrals must be taken between the values of y or x by which the area is bounded.

If we take the first of these expressions, the limiting values of y will be either constants, or functions of x which will be given by the equation of the bounding curve. These being substituted, there will remain for the second integration a function of x alone, the integral of which must be taken between the limits required by the problem.

An expression of the form $A = \int\int dydx$ is called a *double integral.*

153. In like manner, if a solid is referred to rectangular co-ordinates, we may take for the element the infinitely small parallelopiped expressed by the product of the differentials dx, dy, dz. We shall have, therefore,

$$dV = dxdydz;$$

whence

$$V = \int\int\int dxdydz,$$

in which the integral must be taken between proper limits with respect to each of the variables in succession. An expression of this form is called a *triple integral.*

The order of the integrations is altogether arbitrary, and we may employ that which is most convenient. Integrating, first, for example, in respect to z, and supposing the integral to begin when $z = 0$, the expression for the volume becomes

$$V = \int\int dxdyz,$$

in which z will be given in terms of x and y from the equation of the surface. Substituting for z its value, the integration will then be reduced to that of a double integral which will be a function of x and y only.

Ex. 1. The axes of two equal right circular cylinders intersect at right angles; find the volume of the solid, common to both.

Let the intersection of the axes of the cylinders be the origin of co-ordinates, and let their axes be taken for the axes of x and y. The equations for their surfaces, a being the radius of the base of the cylinders, will be

$$x^2 + z^2 = a^2, \; x^2 + y^2 = a^2;$$

and $$V = \int\int\int dxdydz.$$

Integrating first in respect to z, and recollecting that the integral begins when $z = 0$, $\quad V = \int\int dxdyz.$

Substituting next the value of z from the first equation,

$$V = \int\int dxdy(a^2 - x^2)^{\frac{1}{2}}.$$

From the second equation we have $y = (a^2 - x^2)^{\frac{1}{2}}$, which gives for the limits of y, $y = 0$, $y = (a^2 - x^2)^{\frac{1}{2}}$. Integrating between these limits in respect to y, we have next

$$V = \int dx\,(a^2 - x^2).$$

Integrating, finally, this last between $x = 0$, $x = a$, we obtain

$$V = \frac{2a^3}{3}.$$

This is one-eighth of the whole intercepted solid, which will be, therefore, $$\frac{16a^3}{3}.$$

Ex. 2. Find the volume of the elliptic paraboloid, which has for the equation of the surface,

$$\frac{z^2}{a}+\frac{y^2}{b}=2x.$$

We have $\quad V=\int\int z dx dy=\left(\frac{a}{b}\right)^{\frac{1}{2}}\int\int dx dy\,(2bx-y^2)^{\frac{1}{2}}.$

If we make $z=0$ in the equation of the surface we obtain $y^2=2bx$, an equation which establishes a relation between the ordinates x and y, from which we derive for the limits to y, $y=0$, $y=(2bx)^{\frac{1}{2}}$. The integral $\int dy(2bx-y^2)^{\frac{1}{2}}$ expresses a circular area to the radius $(2bx)^{\frac{1}{2}}$ and abscissa y, and which integrated between the limits $y=0$, $y=(2bx)^{\frac{1}{2}}$, (Ex. 2, art. 136) becomes $\frac{1}{2}\pi bx$. Integrating, therefore, the expression for V between $y=0$, $y=(2bx)^{\frac{1}{2}}$, we obtain

$$V=\frac{\pi}{2}(ab)^{\frac{1}{2}}\int x dx=\frac{\pi}{4}(ab)^{\frac{1}{2}}x^2.$$

If we now take this integral from $x=0$ to $x=c$ and multiply by 4, we shall have $\pi(ab)^{\frac{1}{2}}c^2$ for the volume of the paraboloid intercepted between the vertex and a plane parallel to the plane of yz at a distance c.

Ex. 3. Find the volume of the ellipsoid with unequal axes, the equation to which referred to the centre and axes is

$$\frac{x^2}{a^2}+\frac{y^2}{b^2}+\frac{z^2}{c^2}=1,\ \text{or}\ z=\frac{c}{b}\left(\frac{b^2(a^2-x^2)}{a^2}-y^2\right)^{\frac{1}{2}}.$$

Substituting its value for z in the general formula $V=\int\int z dx dy$, and integrating with reference to y between the limits $y=0$, $y=\frac{b}{a}(a^2-x^2)^{\frac{1}{2}}$, we find

$$V=\frac{\pi}{4}\cdot\frac{bc}{a^2}\int(a^2-x^2)\,dx;$$

from which we obtain for the whole volume $V=\frac{4}{3}\pi abc$.

If we make b and c each equal a, we shall have $V=\frac{4}{3}\pi a^3$, the volume of the sphere.

SECTION XV.

APPLICATION TO MECHANICS. VARIED MOTION.

154. Our illustrations of the use of the Calculus as an instrument of investigation, have been derived chiefly from Geometry, and we have seen how important the service it has rendered this science. Problems, which solved by the previous methods, not excepting the more effective ones of Descartes, were tedious and embarrassing, are solved by it with the greatest facility; and others, the solution of which entirely baffled the skill of the ancient geometers, yield at once to its power. Splendid as these triumphs are, they are comparatively insignificant in view of the results it has accomplished in the department of the Physical Sciences, especially in that of Physical Astronomy. Here its powers are most conspicuously displayed and its most wonderful achievements accomplished. We proceed to such illustrations derived from these sources, as the nature of a mere elementary treatise admits, and which may serve to give some idea of the use of the Calculus in this higher and wider field of inquiry.

VARIED MOTION.

155. A force which acts without intermission, and the intensity of which remains the same during the time of its action, is called *a constant accelerating force.*

Let φ represent the velocity generated by the constant force in the unit of time. Then, since the intensity of the force is measured by the velocity it is capable of producing in the unit of time, φ will represent the accelerating force. Let it be supposed that the force has acted for a number of units represented by t, and let v be the velocity generated at the end of the time t; we shall then have $v=\varphi t$, whence

$$\varphi=\frac{v}{t}. \qquad (1)$$

2. A velocity which is constantly changing is called a *variable velocity*. Let v be the variable velocity generated at the end of the time t. To find a measure for this velocity we suppose it to be constant during the next succeeding infinitely small element dt of the time. Let ds, described in the time dt, be the corresponding element of the space, we shall then have

$$v = \frac{ds}{dt}. \qquad (2)$$

3. A force which acts without intermission, and the intensity of which is constantly varying during the time of its action, is called *a variable accelerating force*.

Let φ represent the variable force. Assuming this force to be constant during the infinitely small element dt of the time, the velocity generated in this time will be φdt. But the velocity generated by a constant force in an infinitely small time dt is also infinitely small. Denoting this by dv we shall have $dv = \varphi dt$, whence

$$\varphi = \frac{dv}{dt}. \qquad (3)$$

Differentiating next equation (2), dt being regarded as constant, and substituting in (3) we obtain

$$\varphi = \frac{d^2s}{dt^2}, \qquad (4)$$

or the variable accelerating force is equal to the second differential of the space divided by the square of the differential of the time.

From the equations (2) and (3) a third may be derived which is sometimes useful, viz.

$$ds = \frac{1}{\varphi} v\, dv. \qquad (5)$$

These are the fundamental formulas of varied motion. Let us now apply them to some examples.

Rectilinear Motion. Forces directed to a Centre.

PROBLEM I.

156. Let it be required to determine the motion of a body acted upon by gravity and falling vertically, the variation of gravity and the resistance of the air not being regarded.

In this case the accelerating force, φ, will be constant, and designating it by g, the velocity communicated in a second by gravity at the surface of the earth, we have

$$\frac{d^2s}{dt^2} = g.$$

Multiplying both sides by dt and integrating

$$\frac{ds}{dt} = gt + C = v.$$

But since the velocity is 0, when $t = 0$, the constant C is 0, and we have finally for the velocity

$$v = gt. \qquad (1)$$

2. To find the space, we have $ds = gt\,dt$, from which we obtain

$$s = \tfrac{1}{2} gt^2. \qquad (2)$$

From the equations (1) and (2) we obtain by substitution, $v^2 = 2gs$, or $v = (2gs)^{\frac{1}{2}}$, which gives the velocity acquired by a body falling through a given height.

PROBLEM II.

157. Let it next be required to solve the problem, taking into consideration the variation of gravity without regarding the resistance of the air.

Let S be the centre of the earth, A the point from which the body begins to fall, P its place at any time t from the beginning of the motion. Let $SA = a$, $SP = x$, $SM = r$ the radius of the earth, and $g =$ the force of gravity at the surface of the earth.

S M P A

Then, gravity varying in the inverse ratio of the square of the distance, we have $\varphi : g :: r^2 : x^2$,

whence $$\varphi = \frac{gr^2}{x^2};$$

and giving to φ the negative sign, since the force being directed toward the origin tends to diminish the distance x, we have for the equation of the body's motion

$$\frac{d^2x}{dt^2} = -\frac{gr^2}{x^2}.$$

Multiplying both sides by $2dx$,

$$\frac{2dxd^2x}{dt^2} = -\frac{2gr^2dx}{x^2};$$

whence by integration

$$\frac{dx^2}{dt^2} = 2gr^2\frac{1}{x} + C = v^2.$$

Determining the constant, recollecting that $v = 0$, when $x = a$, we have finally

$$v = (2gr^2)^{\frac{1}{2}}\left(\frac{1}{x} - \frac{1}{a}\right)^{\frac{1}{2}}, \qquad (1)$$

which will give the velocity at any distance x from the centre of the earth, or $a - x$ from the point from which the body begins to fall.

If we suppose $a = \infty$ and $x = r$, we shall have

$$v = (2gr)^{\frac{1}{2}}; \qquad (2)$$

that is, if a body fall from an infinite height to the surface of the earth, the velocity acquired will be finite and equal to $(2gr)^{\frac{1}{2}}$.

If we assume $r = 3965$ miles, and substitute this value in the above equation, we shall have the velocity at the earth's surface equal to 6.9506 miles per second. If then a body be projected upward with a velocity equal to seven miles a second, supposing no resistance from the air, it will not return again to the surface of the earth.

Suppose a body falling from the distance of the sun; to determine its velocity when it arrives at the surface of the earth. Assuming this distance $= 12000$ of the earth's diameters or $a = 24000r$, we shall have by the equation (1)

$$v = \left(2gr \times \frac{23999}{24000}\right)^{\frac{1}{2}}.$$

Performing the calculations we obtain for the velocity sought 6.9505 miles per second.

2. To find the time.

Resuming the expression for the velocity $\frac{dx}{dt}$, affixing the sign —, since x decreases as t increases, and deducing the value of dt, we obtain

$$dt = -\left(\frac{a}{2gr^2}\right)^{\frac{1}{2}} \frac{xdx}{(ax - x^2)^{\frac{1}{2}}};$$

hence by integration

$$t = \left(\frac{a}{2gr^2}\right)^{\frac{1}{2}} [(ax - x^2)^{\frac{1}{2}} - \frac{a}{2} \text{ver-sin}^{-1} \frac{2x}{a} + C].$$

But $t = 0$ when $x = a$; hence $C = \frac{1}{2} a\pi$, and we have

$$t = \left(\frac{a}{2gr^2}\right)^{\frac{1}{2}} [(ax - x^2)^{\frac{1}{2}} - \frac{a}{2} \text{ver-sin}^{-1} \frac{2x}{a} + \frac{a\pi}{2}], \qquad (3)$$

which gives the time for any distance x.

The distance of the moon from the earth is 60 semi-diameters of the earth. Assuming a semi-diameter equal to 3965 miles, the time in which a body would fall from the moon to the earth's surface is 4 days, 19 hours, 46 minutes, 40 seconds.

PROBLEM III.

158. A body descends, within a hollow tube, from a point within the surface of the earth, towards the centre. To determine the circumstances of the motion.

In this case the force will vary as the distance directly. Let S be the centre of the earth, A the point from which the body begins to fall, P its place at the end of any time t from the beginning of the motion, and SM = the radius of the earth.

S P A M

Using the same notation as in the preceding problem, let SA $= a$, SM $= r$, SP $= x$; then $g : \varphi :: r : x$,

whence $$\varphi = \frac{gx}{r};$$

and the equation for the body's motion will be

$$\frac{d^2x}{dt^2} = -\frac{gx}{r};$$

from which we obtain for the velocity at any distance x from the centre

$$v = \left(\frac{g}{r}\right)^{\frac{1}{2}} (a^2 - x^2)^{\frac{1}{2}}. \qquad (1)$$

Here if we make $x = 0$, and $a = r$, we shall have

$$v = (gr)^{\frac{1}{2}},$$

which will be the velocity of the body when it has arrived at the centre from the surface of the earth.

From equation (1) we have, it is evident, $v = 0$ for $x = a$, or $x = -a$. This double value of x shows that when the body has fallen from any point to the centre, it will, in virtue of the velocity thus acquired, rise to an equal distance on the other side, from whence it will again descend, and passing through the centre return to the point from which it first started; and thus continue to vibrate forever.

2. To find the time.

Deducing the value of dt from the equation for the velocity, giving it the sign —, and integrating, we obtain

$$t = \left(\frac{r}{g}\right)^{\frac{1}{2}} \cos^{-1}\frac{x}{a}. \qquad (2)$$

To find the time in which the body will fall to the centre we put $x = 0$, which gives

$$t = \frac{\pi}{2}\left(\frac{r}{g}\right)^{\frac{1}{2}}.$$

This result is remarkable, since it does not involve the distance from which the body begins to fall. Thus the time of arriving at the centre will always be the same from whatever point the motion of the body commences.

Assuming $r = 3965$ miles, the time of falling from the surface, or from any point, to the centre is 21′ 7.3″; and in passing through the whole diameter 42′ 14.6″.

PROBLEM IV.

159. The preceding are only particular cases of the following general problem.

A body P falls from rest from a given point A towards a centre of force S, the force varying as some power of the distance SP. To determine the circumstances of the motion.

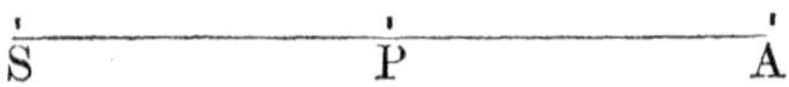

Let SA $= a$, SP $= x$; and let m represent the intensity of the force at the unit of distance. Without proceeding further with the general solution, let the learner now solve the following particular cases.

1. Find the velocity when the force varies directly as the square of the distance.

In this case we shall have for the equation of the body's motion

$$\frac{d^2x}{dt^2} = -mx^2.$$

Proceeding in the same manner as before, we obtain

$$v = \left(\frac{2m}{3}\right)^{\frac{1}{2}} (a^3 - x^3)^{\frac{1}{2}}. \quad \text{Ans.}$$

2. Find the velocity when the force varies in the inverse ratio of the square root of the distance.

Ans. $v = 2m^{\frac{1}{2}} (a^{\frac{1}{2}} - x^{\frac{1}{2}})^{\frac{1}{2}}$.

3. Find the velocity when the force varies inversely as the cube of the distance.

Ans. $v = m^{\frac{1}{2}} \dfrac{(a^2 - x^2)^{\frac{1}{2}}}{ax}$.

4. Find the time when the force varies in the inverse ratio of the cube of the distance.

Ans. $t = \dfrac{a}{m^{\frac{1}{2}}} (a^2 - x^2)^{\frac{1}{2}}$.

PROBLEM V.

160. Let there be next two centres of force, the forces attracting in the inverse ratio of the square of the distance; to determine the circumstances of the motion.

Let A and B be the two forces at the points A and B. Let P, in the right line which joins these points, be the position of the body at the end of the time t, and let D be the point of departure, the motion being from A toward B.

A D P C B

Put $AB = c$, $AD = m$, $AP = x$. Let a^2 represent the intensity of the force A at the unit of distance, and b^2 that of B at the same distance; $\frac{a^2}{x^2}$, and $\frac{b^2}{(c-x)^2}$ will be the intensity of these forces respectively when the body is at P. The accelerating force, it is evident, will be equal to the difference between these two forces, since one tends to increase the distance of the body from A and the other to diminish it. We shall have, therefore, for the equation of the body's motion

$$\frac{d^2x}{dt^2} = \frac{b^2}{(c-x)^2} - \frac{a^2}{x^2}. \qquad (1)$$

Multiplying first by $2dx$, and integrating,

$$\frac{dx^2}{dt^2} = \frac{2b^2}{c-x} + \frac{2a^2}{x} + C = v^2. \quad (2)$$

To determine the constant, let $k =$ the velocity of the body on its departure from the point D. We shall have, substituting for x in (2) its value m,

$$k^2 = \frac{2b^2}{c-m} + \frac{2a^2}{m} + C. \qquad (3)$$

Deducing the value of C and substituting in (2)

$$v^2 = k^2 + 2b^2\left(\frac{1}{c-x} - \frac{1}{c-m}\right) - 2a^2\left(\frac{1}{m} - \frac{1}{x}\right); \quad (4)$$

from which the velocity may be found for any position of the body between the points A and B.

There will be, it is evident, on the line AB which joins A and B, a point C in which the two forces are equal; so that if the body should be placed at this point, or should arrive at it without any velocity, it will remain in equilibrium.

Let h be the distance AC of this point from A; then

$$\frac{b^2}{(c-h)^2}=\frac{a^2}{h^2};$$

from which we deduce two values of h, one of which belongs to the point C between A and B, and the other to a point C′, situated on AB produced on the side of the least force. The first of these two values is

$$h=\frac{ac}{a+b}.$$

Let v' be the least initial velocity which must be impressed upon the body in order that it may reach the point C with its velocity reduced to 0; then

$$k=v',\ x=h=\frac{ac}{a+b},\ \text{and}\ v=0;$$

whence by substitution in (4) we obtain

$$v'^2=\frac{2b^2}{c-m}+\frac{2a^2}{m}-\frac{2(a+b)^2}{c}. \quad (5)$$

If the initial velocity k is less than v', the movable will return back toward A; if greater, it will pass through C and move on towards B. If k is equal to v', the body will move on to C and there remain suspended between A and B.

If A and B are the centres of two homogeneous spheres, the forces may be regarded as the attractions of the two spheres respectively, and the intensities a^2 and b^2 will be to each other as the masses of the spheres. Thus, let B be the centre of the earth, A that of the moon; the mass of the latter being assumed as one seventy-fifth of the former, we shall have

$$a^2=\frac{b^2}{75},$$

whence

$$h=\frac{c}{1+\sqrt{75}}=0.10352c.$$

The point of equal attraction between the earth and moon is, therefore, very nearly *one-tenth* of the distance between them reckoned from the moon.

PROBLEM VI.

161. Let the body be acted upon by an attracting and repulsive force.

Suppose the body placed at A to be acted upon by an attracting force tending to make it move from A towards B with a motion uniformly varied, and a repulsive force varying inversely as the distance from the point B and tending to push the body from B towards D.

Let AB $= a$, and N being the place of the body at the end of the time t, let AN $= x$. Let $m =$ the intensity of the repulsive force at the unit of distance, and $g =$ the uniform accelerating force. The equation of the body's motion will be

$$\frac{d^2x}{dt^2} = \frac{m}{a+x} - g;$$

and we obtain finally for v^2

$$v^2 = 2m \log(a+x) - 2m \log a - 2gx,$$

which will give the velocity when the body has passed over any space x.

The formula for v would find its application in the case of a piston placed in a vertical cylinder open at top, and in which it moves freely. The piston is urged downward by gravity with a constant accelerating force, and pushed back by the elastic force of the air beneath, varying in intensity in proportion inversely to the distance of the piston from the bottom of the cylinder.

Forces directed in any manner in the same plane. Curvilinear motion.

162. In the preceding examples the directions of the forces have all been in the same right line. But the forces may be directed in any manner with respect to each other. We shall take the most simple case, that in which the directions of the forces are all in the same plane. We have then the following general problem, viz.

To find the equations of motion of a point moving in a plane and acted upon by any forces in that plane.

Let the forces be resolved each into two others parallel to two rectangular axes taken in the plane of the forces. Let X represent the aggregate of the resolved forces parallel to the axis of x, and Y the aggregate of those parallel to the axis of y. Let x and y be the co-ordinates of the position of the point, and t the time reckoned from a given epoch. For the forces in x and y respectively we have, art. 155,

$$\frac{d^2x}{dt^2}=X, \quad \frac{d^2y}{dt^2}=Y, \quad (1)$$

in which X and Y will be positive or negative according as they tend to increase or decrease the values of x and y respectively.

Let ds be the element of the path of the point. Then $\frac{ds}{dt}$ will be the velocity of the body in its path. Let α and β be the angles which the element ds makes with the axes of x and y. We shall then have

$$\frac{ds}{dt}\cos\alpha=\frac{dx}{dt}; \quad \frac{ds}{dt}\cos\beta=\frac{dy}{dt}. \quad (2)$$

These are the general equations by means of which the motion of the point is determined. The motion may be either rectilineal or curvilinear.

163. We shall take a particular case of the two following general problems. 1°. Given the forces which act upon a body to determine its path or trajectory. 2°. The motion being known, to determine the forces that will cause the body to describe a given trajectory.

I. Suppose a repulsive force to vary inversely as the cube of the distance from a given plane, and let a body be projected from a point at a given distance from the plane, and with a given velocity in a direction parallel to the plane; to determine the trajectory.

Let AX, AY, be two rectangular axes, one of which AX is in the given plane, and the other AY perpendicular to it, and passing through the point N, from which the body is projected in a line NS parallel to AX. The trajectory, it is evident, will be convex to the given plane, and will be in the plane of the axes AX, AY. Let M be the place of the body at the end of the time t; let $\text{AP} = x$, $\text{PM} = y$, and $a =$ the distance of the point of projection of the body from the given plane. Let m represent the intensity of the repulsive force at the unit of distance, and v the given velocity of projection. We shall have for the equations of the body's motion

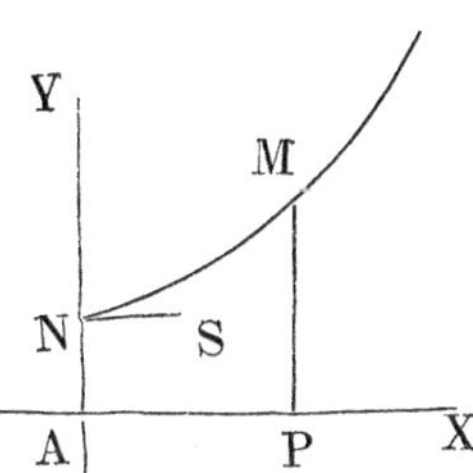

$$x = vt, \text{ or } dx = vdt, \qquad (1)$$

$$\frac{d^2y}{dt^2} = \frac{m}{y^3}. \qquad (2)$$

Integrating equation (2) by the usual process, recollecting that the velocity in the direction of the axis of y is 0 when $y = a$, we obtain

$$\frac{dy^2}{dt^2} = m\left(\frac{1}{a^2} - \frac{1}{y^2}\right). \qquad (3)$$

Eliminating next dt by means of the equations (1) and (3), we have, m being 1,

$$\frac{aydy}{(y^2-a^2)^{\frac{1}{2}}} = \frac{dx}{v}. \qquad (4)$$

Integrating this, observing that for $x = 0$, $y = a$, we obtain

$$a\,(y^2 - a^2)^{\frac{1}{2}} = \frac{x}{v};$$

an equation which contains the variables x and y only, and which determine the trajectory.

Deducing the value of y^2 from this equation we have

$$y^2 = \frac{x^2}{a^2v^2} + a^2 = \frac{1}{a^2v^2}(x^2 + a^4v^2),$$

or

$$y^2 = \frac{a^2}{a^4v^2}(x^2 + a^4v^2),$$

the equation of the trajectory. The trajectory is, therefore, an *Hyperbola* whose semi-axes are a^2v and a respectively.

From what has been done we see that if the functions X and Y in equations (1) art. 162, are given by the conditions of the problem, and the integrals of these equations can be taken, they will contain the three variables x, y and t only; and if t can be eliminated, we shall obtain an equation in x and y only; which, expressing the relation between x and y, will give the equation of the trajectory or path described by the body.

II. To find the forces which must act upon a point, so that it may describe the arc of a parabola with a uniform motion.

Let s represent the curve, x and y the co-ordinates to any point, and $y^2 = 4ax$ be the equation to the parabola. Then, since the velocity is constant, we shall have

$$\frac{ds}{dt} = C, \text{ a constant quantity;}$$

whence

$$C^2 = \frac{ds^2}{dt^2} = \frac{dx^2}{dt^2} + \frac{dy^2}{dt^2}.$$

From the given equation $y^2 = 4ax$, we have

$$\frac{dy}{dx} = \left(\frac{a}{x}\right)^{\frac{1}{2}};$$

hence

$$C^2 = \frac{dx^2}{dt^2}\left(1 + \frac{a}{x}\right), \text{ and } \frac{dx^2}{dt^2} = \frac{C^2}{1 + \frac{a}{x}}.$$

Differentiating this last and reducing, we obtain

$$\frac{d^2x}{dt^2} = \frac{C^2a}{2(a+x)^2},$$

which gives the force parallel to the axis of x.

In a manner altogether similar we obtain

$$\frac{d^2y}{dt^2} = -\frac{C^2y}{4(a+x)^2},$$

for the force parallel to the axis of y. The negative sign shows that the force tends toward the axis of x.

Combining these two forces we shall have the whole force which acts upon the point in any position in its path.

Motion along Curve Lines.

164. The preceding formulas for curvilinear motion apply equally to motion along curve lines, the reaction of the curve in the direction of its normal being regarded as one of the forces acting upon the body, and being resolved, with the others, in the direction of the co-ordinates. It will, however, be more simple in the problems we propose to solve, to consider only the forces along the curve.

Let AMB be the curve along which the body moves, and BZ the vertical axis to which the curve is referred. Let M be the point at which the body has arrived at the end of the time t. Let s be the arc BM reckoned from the lowest point B. Let ME be one of the infinitely small sides of which the curve AMB is supposed to be composed; then ME $= ds$; and dt will be the time in which ds is described. Draw NO perpendicular to ME, meeting the vertical MO in the point O. Then MO representing the vertical force of gravity, this force may be decomposed into two forces NO and MN, one of which is perpendicular to, and the other acting along the curve. The perpendicular force NO is destroyed by the reaction of the curve, and the only force which will cause the body to move along the curve is MN. Draw the perpendicular MP meeting the vertical ER in R; then by similar triangles

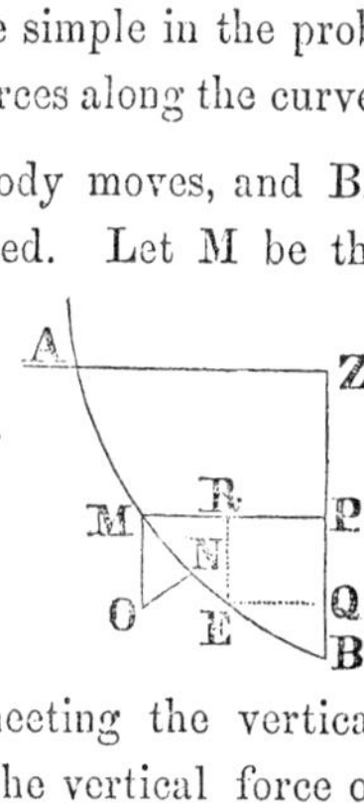

MO : MN : : ME : ER.

Let BP $= x$, PM $= y$; then ER $=$ PQ $= dx$, and g being the

vertical force of gravity, we have for the accelerating force along the curve

$$\varphi = \text{MN} = \frac{\text{MO} \times \text{ER}}{\text{ME}} = g\,\frac{dx}{ds},$$

whence by substitution in formula (5) art. 155, we have

$$vdv = -gdx.$$

We place the sign — before gdx because the arc decreases when the velocity increases.

Integrating this equation we obtain

$$v^2 = 2C - 2gx.$$

To determine the constant, let it be supposed that the point from which the body begins to fall is A. Draw AZ perpendicular to BZ; then BZ will be the vertical height through which the body falls in its descent from A to B. Let BZ $= h$, then if $x = h$, $v = 0$, and we have $C = gh$; whence

$$v^2 = 2gh - 2gx = 2g(h - x) \qquad (1)$$
$$= 2g \times \text{ZP}.$$

Now if a heavy body fall through ZP, the square of its velocity V^2, on arriving at P will be

$$\text{V}^2 = 2g \times \text{ZP}.$$

Therefore, *when a heavy body descends along any curve line, it has, at any point whatever, the velocity it would have acquired by falling freely through a space of the same perpendicular elevation.*

It is evident, moreover, that the velocity which a body successively acquires by gravity in descending along a curve is entirely independent of the nature of the curve. It will be the same for all curves whatever.

Substituting in the equation $v = \frac{ds}{dt}$ for v its value from (1), and affixing to it the negative sign, we have

$$dt = \frac{-ds}{(2g)^{\frac{1}{2}}(h - x)^{\frac{1}{2}}}.$$

To find, therefore, the time employed in describing any arc BM of the curve, we must substitute for ds its value in terms of x and dx derived from the equation to the curve, and then integrate.

PROB. Let it be required to find the time in which a heavy body will fall through any arc BM of the cycloid, whose base is horizontal and vertex downward.

We have, art. 147, for the length s of any arc of the cycloid reckoning from the vertex B,

$$s = 2(2rx)^{\frac{1}{2}},$$

r being the radius of the generating circle; or putting a for the diameter of the generating circle

$$s = 2a^{\frac{1}{2}}x^{\frac{1}{2}};$$

whence, substituting in the formula for dt and integrating

$$t = -\left(\frac{a}{2g}\right)^{\frac{1}{2}} \text{ver-sin}^{-1}\frac{2x}{h} + C.$$

Taking this between the limits $x = h$, $x = 0$, we obtain

$$t = \pi\left(\frac{a}{2g}\right)^{\frac{1}{2}}.$$

This result, it is evident, is entirely independent of the length of the arc s. If a body, therefore, descend along an arc of a cycloid having its base horizontal and vertex downward, *the time of reaching the lowest point will be the same whatever the length of the arc.*

Since the body having fallen to B will, with the velocity acquired, ascend in the same time to the same height on the other side of the curve as that from which it has fallen, the whole time of the oscillation of a body moving on a cycloid will be $2\pi\left(\frac{a}{2g}\right)^{\frac{1}{2}}$, or $\pi\left(\frac{2a}{g}\right)^{\frac{1}{2}}$. And since the times of these oscillations will be equal, whatever the lengths of the arcs, the cycloid is said to be *tautochronous.*

The property of tautochronism, first discovered by Huygens, led to the application of a pendulum vibrating in a cycloid to regulate the motion of a clock.

Since the cycloid is at once its own involute and evolute, a pendulum may be made to vibrate in a cycloid in the following manner.

Let two equal semi-cycloids CA, CB be placed in the same vertical plane with their bases horizontal and vertices downward, and let them be joined in a common point C at the extremity of the bases, their convexities being turned toward each other. Let a ball be suspended by a fine thread from C, the distance between the centre of the ball and C being taken equal to the curve of either of the semi-cycloids. If the ball be now made to vibrate between the two semi-cycloidial arcs, the string will apply itself successively from one to the other, and the ball will thus be made to vibrate in a cycloid. The impulse by which the vibrations of the pendulum are maintained should be communicated to it, it is obvious, when at the lowest point of its motion.

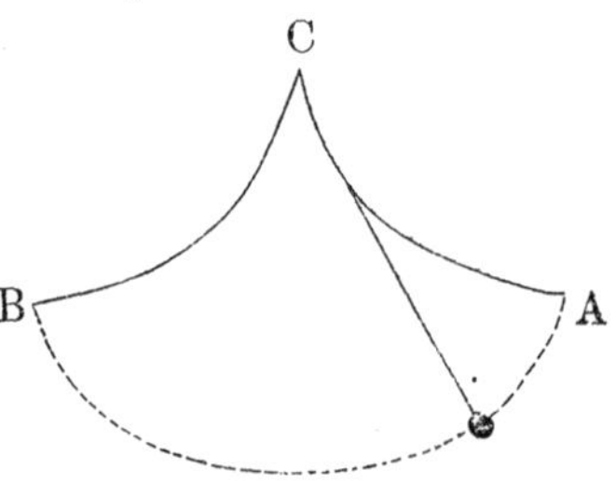

165. A curve revolves uniformly about a fixed axis; suppose a body P descending along the curve by the action of gravity; to determine the velocity of the body at any point.

Let AC be the curve, AB the axis about which it revolves and which we suppose to be vertical. Let the arc AP $= s$; draw PM and CB perpendicular to AB, and let AM $= x$, PM $= y$, AB $= h$, CB $= a$. Produce PM to R, and let PR represent the centrifugal force developed at P by the rotation of the curve; and let this force be resolved into two, one

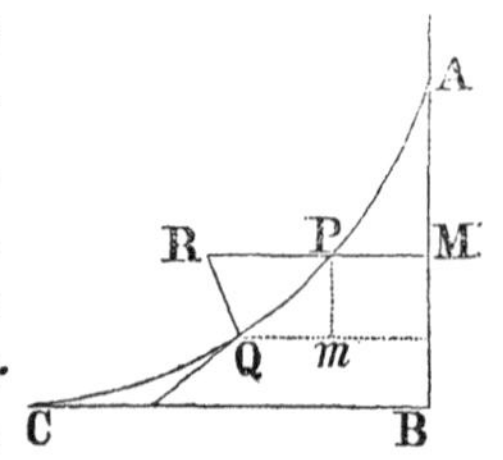

RQ perpendicular to the curve, and the other PQ in the direction of the tangent at P. Then

$$PQ = PR \cos RPQ = PR \,.\, \frac{dy}{ds}.$$

Let V represent the velocity at the point C of the curve; that at P will be

$$\frac{y}{a} V.$$

Then, since the centrifugal force is equal to the square of the velocity divided by the radius, we have

$$PR = \frac{y^2}{a^2} V^2 \frac{1}{y} = \frac{V^2}{a^2} y.$$

We shall have, therefore, for the accelerating force resulting from the rotation of the curve

$$\varphi = PQ = \frac{V^2}{a^2} \frac{ydy}{ds}.$$

Again, the accelerating force resulting from gravity is

$$\varphi' = g \frac{dx}{ds};$$

hence we have for the whole accelerating force

$$\varphi + \varphi' = \frac{V^2}{a^2} \frac{ydy}{ds} + g \frac{dx}{ds}.$$

Substituting next in formula (5), art. 155, and integrating, we obtain

$$v^2 = \frac{V^2}{a^2} y^2 + 2gx. \qquad (1)$$

Determining next the relation of the variables x and y for each point of the curve, and eliminating one of them, we shall obtain an expression by which the velocity at any point of the curve may be found.

Prob. Let the curve be the quadrant of a circle convex to the axis of rotation which also touches it.

1. To find the velocity at any point,

Let AC in the preceding figure be the curve, O the centre, AB the axis of rotation. Let the centre O be the origin of the co-ordinates; then a being the radius of the circle, we have

$$x^2 = a^2 - (a - y)^2,$$

or

$$x = (2ay - y^2)^{\frac{1}{2}};$$

whence by substitution in (1)

$$v^2 = \frac{V^2}{a^2} y^2 + 2g (2ay - y^2)^{\frac{1}{2}}.$$

Let t denote the time in which the point C will describe the circumference of the circle of which $BC = a$ is the radius, we shall have

$$V = \frac{2\pi a}{t};$$

whence

$$v^2 = \frac{4\pi^2}{t^2} y^2 + 2g (2ay - y^2)^{\frac{1}{2}},$$

which will give the velocity at any point.

SECTION XVI.

APPLICATION TO MECHANICS CONTINUED. EQUILIBRIUM. CENTRE OF GRAVITY. HYDRODYNAMICS.

166. In order that a material point or body may remain at rest in the same position, the resultant of all the forces which act upon it must be equal to 0.

Let the directions of all the forces which act upon the body be in the same plane; and let the aggregate of the forces resolved in the directions of two rectangular axes be represented by X and Y respectively. The equations of equilibrium will then be

$$X = 0, \quad Y = 0.$$

Let it be proposed to find the conditions of equilibrium when a body is supported on a curve, the curve being in a vertical plane, and referred to rectangular axes.

Let $AM = x$, $MP = y$, be the co-ordinates of the body at the point P of the curve, and let the arc $BP = s$. Let the forces which act upon the body, resolved in the directions parallel to the axes of x and y, be represented by X and Y respectively; X and Y being regarded as pesitive when they tend to increase x and y, and negative when they tend to diminish them. Let R be the reaction of the curve in the direction of the normal, or which is the same thing, the pressure of the body upon the curve.

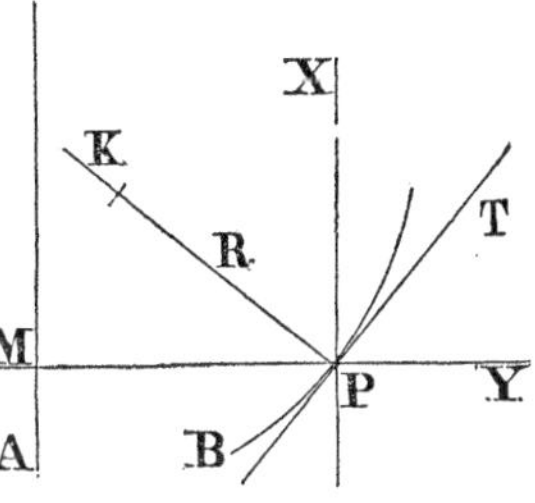

Resolving R in the directions parallel to x and y, we have for the resolved part in the direction PX

$$R \cos RPX = R \sin XPT = R\frac{dy}{ds};$$

and for the part in the direction PY

$$R \cos RPY = - R \cos RPM = - R\frac{dx}{ds}.$$

By the conditions of equilibrium we must have then

$$X + R\frac{dy}{ds} = 0, \qquad Y - R\frac{dx}{ds} = 0.$$

The first of these equations is evidently equivalent to

$$X + R\frac{dy\,dx}{dx\,ds} = 0;$$

and if we multiply the second by $\frac{dy}{dx}$, add it to this last, and reduce,

we obtain $$Xdx + Ydy = 0; \qquad (1)$$

which is the equation of equilibrium sought. If now the equation to the curve is given, we shall know the relation between the forces X and Y, by means of which and the equation to the curve, the

17*

particular values of x and y may be found, for which the body will be supported on the curve, or be in equilibrium.

So far as the conditions of equilibrium are concerned, it is immaterial, it will be observed, whether the body rests upon a material surface BP, or is suspended by a string KP which confines it to this surface. The tension of the string in the direction of the normal, supplies in this case the reaction of the curve.

Prob. A body P suspended by a string PC without weight, is compelled to remain somewhere on the circle of which PC is the radius. Suppose the body to be acted upon by gravity, and by a repulsive force, acting from the lowest point A of the circle, in the inverse ratio of the square of the distance; to determine the position of equilibrium.

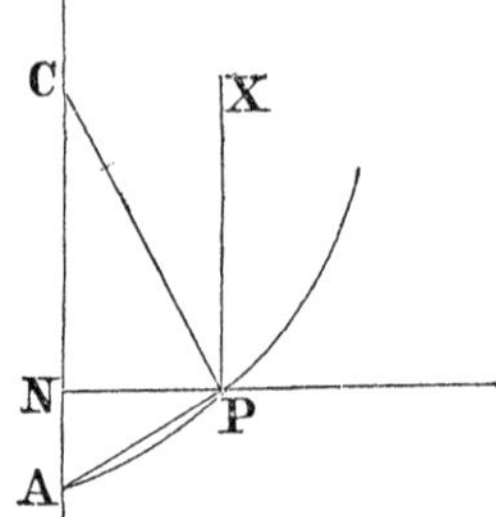

Let CA $=$ CP $= a$, AP $= p$, AN $= x$, NP $= y$. Let $m =$ the repulsive force at the unit of distance; at the distance p it will be $\frac{m}{p^2}$. Resolving this force in the directions of x and y, we have for the resolved parts

$$\frac{m}{p^2}\frac{x}{p}, \text{ and } \frac{m}{p^2}\frac{y}{p};$$

whence, putting g for the force of gravity, and substituting in the general equation of equilibrium, (1), we obtain

$$\left(\frac{mx}{p^3} - g\right) dx + \frac{my}{p^3} dy = 0.$$

Eliminating dy by means of the equation to the circle, $y^2 = 2ax - x^2$, and reducing, we obtain

$$p = \left(\frac{ma}{g}\right)^{\frac{1}{3}};$$

from which the position of the body when in equilibrium is known.

CENTRE OF GRAVITY.

167. Prob. I. *To find the centre of gravity of a plane surface bounded by a curve line.*

Let APM be the surface. We suppose this surface divided into an infinite number of infinitely small trapezoids MPpm. The centre of gravity of the infinitely small trapezoid MPpm will be at F, the middle point of the line PM. Let OX, OY be any system of rectangular axes; and draw FK perpendicular to OY. The moment of the trapezoid MPpm with reference to the axis OY, will be FK $\times$ MPpm. Let G be the centre of gravity of the surface APM; then the distance GG′ of G from the axis OY will, by Mechanics, be equal to the sum of the moments of the infinitely small trapezoids MPpm divided by the sum of these trapezoids or the surface APM. Putting $m =$ to this surface and regarding OX and OY as the axes of x and y, we shall have FK $= x$,

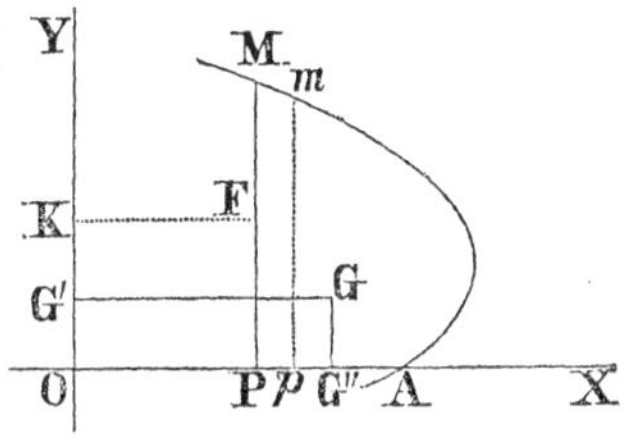

and $$\text{FK} \times \text{MP}pm = xdm;$$

whence $$\text{GG}' = \frac{\int xdm}{m}. \qquad (1)$$

By a process altogether similar we obtain for the distance of the centre of gravity G from the axis OX

$$\text{GG}'' = \frac{\int \frac{1}{2} ydm}{m}. \qquad (2)$$

If the second of these equations is equal to 0, the centre of gravity will be on the axis OX; if the first, it will be on the axis OY; if both at the same time are equal to 0, it will be at the origin 0.

Ex. 1. To find the centre of gravity of a segment of a circle.

Let O be the centre of the circle, MAM′ the segment whose centre of gravity is sought; and let the axes to which the centre of gravity is referred be OA, OY. Then, since the area is symmetrical on both sides the axis OA, the centre of gravity G will fall upon OA, and we shall have only to calculate its distance OG from the centre O. In this case, a being the radius of the circle, A the origin, we have in the formula (1), $x = a - x$, $dm = 2ydx$, and $m = 2\text{APM}$;

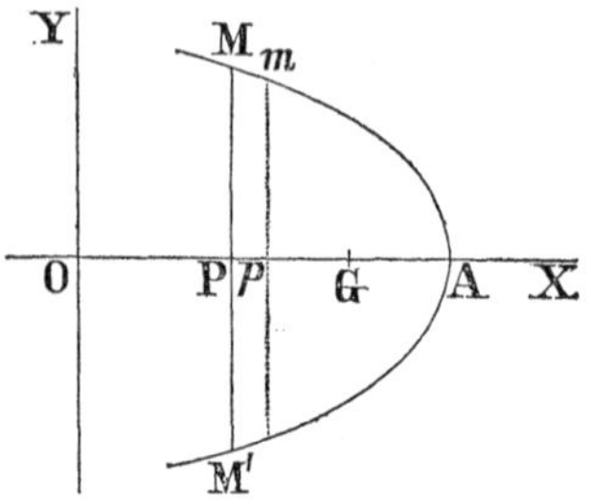

whence $$GG' = \frac{\int (a - x) 2ydx}{2\text{APM}} = \frac{\int (a - x)\, ydx}{\text{APM}}.$$

Substituting for y its value from the equation to the circle $y^2 = 2ax - x^2$, integrating, and determining the constant by the hypothesis $x = 0$, we obtain

$$OG = \frac{(2ax - x^2)^{\frac{3}{2}}}{3\,\text{APM}} = \frac{\text{PM}^3}{3\,\text{APM}}.$$

If MAM′ be a semicircle, then $OG = \frac{4}{3\pi}\text{AO}$.

Ex. 2. To find the centre of gravity of the common parabola whose equation is $y^2 = px$.

Ans. Three-fifths of the distance from the vertex to the chord which limits the surface.

PROB. II. *To find the centre of gravity of any arc of a curve.*

Let AM in the preceding figure be the arc. The distance of the centre of gravity of this arc from the axes OX, OY respectively, will be equal to the sum of the moments with reference to these axes of the infinitely small arcs Mm, divided by the sum of the arcs or the whole arc AM. It will be sufficient to put, therefore, in the formulas (1) and (2) already obtained, $m =$ the arc AM, and dm = the differential of this arc.

Ex. 1. To find the centre of gravity of the arc of a circle.

Let O be the centre of the circle, MAM′ the arc symmetrically situated on each side the axis OX. Then

$$GG' = \frac{2\int xdm}{MAM'},$$

in which, A being the origin of the co-ordinates, $x = a - x$. Substituting, putting for dy^2 its value, and integrating we find

$$\int xdm = \int (a - x)\,(dx^2 + dy^2)^{\frac{1}{2}} = a(2ax - x^2)^{\frac{1}{2}},$$

the constant being 0. We have, therefore,

$$OG = \frac{a2\,(2ax - x^2)^{\frac{1}{2}}}{MAM'};$$

from which it follows, that the distance of the centre of gravity of any arc of a circle from the centre of this circle is a fourth proportional to the length of the arc, its chord, and radius.

For the semicircle we have $OG = \dfrac{2a^2}{\pi}$.

Ex. 2. Let the curve be the cycloid.

Ans. The centre of gravity is on the axis, at the distance of one-third of this line from the vertex.

Prob. III. *To find the centre of gravity of a solid of revolution.*

In the preceding figure let AO, the axis of x, be the axis of revolution. The solid being symmetrical in respect to its axis AO, the centre of gravity will be in this line. Putting m for the volume of the solid, we have for its differential, art. 52, $dm = \pi y^2 dx$. Hence

$$GG' = \frac{\int \pi y^2 x dx}{m}.$$

Ex. 1. To find the centre of gravity of any segment of the sphere.

Supposing the origin O at the vertex A, recollecting that $m = \int \pi y^2 dx$, and substituting for y^2 its value from the equation $y^2 = 2ax - x^2$, we have

$$AG = \frac{\int \pi (2ax - x^2)\, xdx}{\int \pi\,(2ax - x^2)\,dx};$$

whence we deduce $$AG = \frac{8ax - 3x^2}{4(3a - x)}.$$

Putting $x = a$, we have for the whole hemisphere $AG = \frac{5a}{8}$.

That is, the distance of the centre of gravity of a hemisphere from the vertex, is five-eighths the radius of the generating circle.

Ex. 2. To find the centre of gravity of a paraboloid, the generating curve being $y^2 = 2px$.

Ans. Two-thirds the altitude from the vertex.

PROB. IV. *To find the centre of gravity of a surface of revolution.*

The same things being supposed as in the preceding problem, we have for the element of the surface, art. 58, $dm = 2\pi y(dx^2 + dy^2)^{\frac{1}{2}}$; whence, the origin being taken at the vertex,

$$AG = \frac{\int 2\pi y(dx^2 + dy^2)^{\frac{1}{2}}x}{m}.$$

Ex. 1. To find the centre of gravity of the surface of a spherical segment.

Substituting for m we have

$$AG = \frac{\int 2\pi y(dx^2 + dy^2)^{\frac{1}{2}}x}{\int 2\pi y(dx^2 + dy^2)^{\frac{1}{2}}}.$$

Substituting next for dy^2 its value from the equation to the generating circle, reducing and integrating, we obtain

$$AG = \tfrac{1}{2}x.$$

That is, the distance of the centre of gravity of the surface of a spherical segment from the vertex, is one-half the altitude of the segment.

Ex. 2. To find the centre of gravity of the surface of a cone.

Ans. Two-thirds the distance from the vertex.

HYDRODYNAMICS.

I. *Pressure of Fluids*,

168. To determine the pressure of a fluid upon the sides and bottom of a vessel which contains it, we suppose the fluid divided into an indefinite number of infinitely thin parallel segments. The pressure upon a point K of any one of these elementary segments QPqp is equal to the weight of the vertical filament IK of the fluid. And the pressure upon the vertical section Qq of the segment, will be equal to the weight of IK taken as many times as there are points K in Qq.

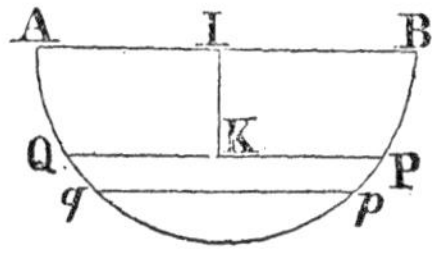

Let IK $= x$, $s =$ the density of the fluid, and $z =$ the vertical section AQ of the vessel, corresponding to the depth IK. The weight of the filament IK will, it is evident, be equal to $\int sdx$, and the pressure upon Qq will be Q$q\int sdx$. Thus we shall have for the element of the pressure P upon any vertical section AQ of the side of the vessel,

$$dP = dz \int sdx.$$

Let L be the length of the elementary segment; then the element of pressure for the surface will be

$$dP = Ldz \int sdx. \qquad (1)$$

Prob. 1. To find the whole pressure upon the sides of a cubical vessel filled with a heavy fluid, the density of the fluid being regarded as 1.

Let $a =$ the length of a side of the vessel; then $L = 4a$, and

$$P = \int 4axdx,$$
$$= 2ax^2 + C.$$

Taking this from $x = 0$ to $x = a$, the whole pressure upon the sides of the vessel will be $2a^3$. The pressure upon the bottom, it will be easily seen, is a^3. Thus the pressure upon the sides of the vessel is twice that on the bottom.

169. A segment of a solid of revolution is exactly immersed in water with its vertex downward. To determine the pressure upon the surface, the density of water being regarded as unity.

Let $h =$ the altitude of the segment, x any abscissa reckoning from the vertex, and y the corresponding ordinate. Then in equation (1) $L = 2\pi y$, $dz = (dx^2 + dy^2)^{\frac{1}{2}}$, and $dx = d(h - x)$; we shall have, therefore,

$$d\,P = 2\pi y\,(dx^2 + dy^2)^{\frac{1}{2}}\,(h - x). \qquad (2)$$

Prob. Let the solid immersed be the segment of a sphere with its vertex downward.

By substitution in (2), r being the radius of the sphere,

$$d\,P = 2\pi r\,(h - x)\,dx;$$

whence $$P = 2\pi r\,(hx - \tfrac{1}{2}x^2).$$

Putting $x = h$, we have for the whole pressure

$$P = \pi r h^2.$$

For a hemisphere the pressure will be πr^3. And for the whole sphere it will be $4\pi r^3$.

170. Let it be supposed next that the density varies according to any law; to determine the pressure.

In this case we suppose the elementary strata to be of different densities, the densities varying according to the given law. Let S represent the variable density at any depth x below the surface. The weight of the filament IK, in the preceding figure, will be $\int S dx$, and the expression for the element becomes

$$d\,P = L\,dz \int S dx. \qquad (3)$$

Prob. I. A cylindrical vessel is filled with a heavy fluid, the density of which varies as the depth; to determine the pressure which any part of the side of the vessel has to sustain.

Let the part required be a vertical section of the side. Let h be the length of the section or height of the vessel, S' the density of the fluid at the base, and x any variable depth; then

$$S = \frac{S'x}{h};$$

and $$P = \int dx \int \frac{S'x\,dx}{h} = \frac{S'x^3}{6h} + C;$$

from which we obtain for the required pressure $\frac{1}{6}S'h^2$.

PROB. II. A conical vessel, resting on its base, is filled with a fluid, whose density varies as the depth; to determine the pressure upon the side.

Let h be the altitude of the cone, r the radius of the base, x any depth from the vertex, y the radius of the section at the depth x; then, S′ being the density at the base, we have for the element of the pressure

$$d\text{P} = 2\pi y(dx^2 + dy^2)^{\frac{1}{2}} \int \text{S}' \frac{xdx}{h}.$$

But $x : y :: h : r$; hence by substitution and integration

$$\text{P} = \pi \frac{\text{S}'r}{h^2} \frac{(h^2 + r^2)^{\frac{1}{2}}}{h} \frac{x^4}{4} + C;$$

and for the pressure sought

$$\text{P} = \frac{1}{4} \pi \text{S}' r (h^2 + r^2)^{\frac{1}{2}} h.$$

II. *Discharge of Fluids through Apertures.*

171. If water be discharged from an orifice in the bottom or side of a vessel which is kept constantly full, the velocity will be equal to that acquired by a heavy body in falling through a space equal to the depth of the orifice below the line of level of the fluid. Putting x for this depth, the velocity will be equal to $(2gx)^{\frac{1}{2}}$.

Let A $=$ the area of the orifice; the quantity Q of the fluid, discharged in a determinate time t, will be

$$\text{Q} = \text{A}t\,(2gx)^{\frac{1}{2}}. \quad (1)$$

If the vessel be suffered to empty itself, the velocity of the descending surface, and that of the fluid discharged, will be uniformly retarded. It is proposed to determine the time in which a vessel of any form will empty itself. Let $\text{A}' =$ the area of the descending surface, A being that of the orifice and x the depth of the fluid at the end of any time t. In the infinitely small instant of time dt, the quantity of fluid discharged through the orifice will be $\text{A}dt\,(2gx)^{\frac{1}{2}}$. But in the same time the upper surface of the fluid has descended through the space dx; and the vessel has lost a quantity of fluid equal to $A'dx$. We shall have, therefore,

$$A'dx = A dt\,(2gx)^{\frac{1}{2}};$$

whence
$$dt = \frac{A'dx}{A\,(2gx)^{\frac{1}{2}}}; \qquad (1)$$

from which t may be found by integration, A′ being either constant, or capable, from the form of the vessel, of being determined in functions of x.

PROB. I. Let it be required to find the time of emptying a cylinder or prism, through an orifice in the bottom.

In this case A′ will be constant, and we have

$$dt = -\frac{A'}{A\,(2g)^{\frac{1}{2}}} \cdot \frac{dx}{x^{\frac{1}{2}}},$$

the sign — being given to x, since x decreases as t increases.

Integrating, and determining the constant by recollecting that when $x = h$, $t = 0$, we obtain

$$t = \frac{2\,A'}{A\,(2g)^{\frac{1}{2}}}\,(h^{\frac{1}{2}} - x^{\frac{1}{2}}).$$

To find the time of completely emptying the vessel we have only to put $x = 0$, which gives

$$t = \frac{A'(2h)^{\frac{1}{2}}}{A g^{\frac{1}{2}}}. \qquad (1)$$

But from what has been shown above, we have

$$Q, \text{ or } A'h = At\,(2gh)^{\frac{1}{2}};$$

whence
$$t = \frac{A'\,(2h)^{\frac{1}{2}}}{2\,A g^{\frac{1}{2}}}. \qquad (2)$$

Comparing these results we see that when a vessel is left to empty itself, *the time employed is just double that required to discharge the same quantity when the vessel is kept full.*

PROB. II. To determine the time in which a given sphere will empty itself through an orifice at the lowest point.

Let $r =$ the radius of the sphere, x any depth from the bottom; then $A' = \pi(2rx - x^2)$, and

$$dt = -\frac{\pi}{A(2g)^{\frac{1}{2}}} \cdot \frac{(2rx - x^2)\,dx}{x^{\frac{1}{2}}};$$

whence, by integration we obtain for the whole time of emptying the vessel,

$$t = \frac{16\pi r^{\frac{5}{2}}}{15Ag^{\frac{1}{2}}}. \quad \text{Ans.}$$

Ex. 2. To find the time of emptying a given cone through an orifice in the vertex, the base of the cone being horizontal and its vertex downward.

Let $r =$ the radius of the base, h the altitude of the cone.

$$\text{Ans.} \quad \frac{2\pi r^2 h^{\frac{1}{2}}}{5A(2g)^{\frac{1}{2}}}.$$

SECTION XVII.

METHOD OF VARIATIONS.

172. In the problems already solved by aid of the auxiliary quantities furnished by the Calculus, the dependence of the variables has remained the same throughout the entire process for their solution. But there are other problems in which this is not the case; and the solution of which becoming in consequence more difficult, requires additional aids in order to its accomplishment.

Let it be required, for example, to trace upon a plane, between two points A and D, a curve, of a given length, such that, AK being the axis and DK any ordinate perpendicular to this axis, the area comprised between the curve and the ordinates AK, KD may be the greatest possible.

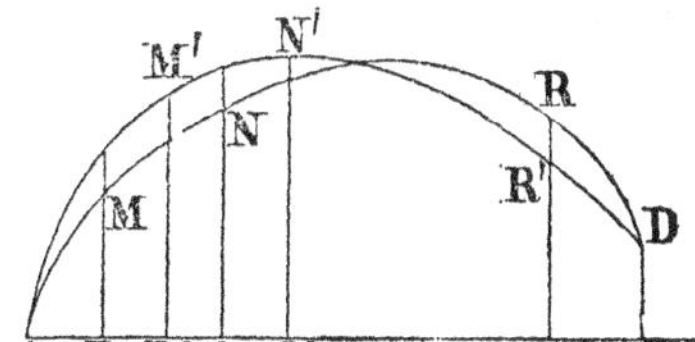

This problem though comprised in the general theory of maxima and minima, is materially different from those we have already considered. In the latter the relation between the variables, given by the conditions, remains the same throughout. All that we are required to do is, to determine the particular values which must be assigned to the variables, in order that the maximum or minimum sought may exist. But in the problem now proposed, the relation itself between the variables, instead of being given, is the thing sought. The question is, among all the possible relations which may exist between the variables, to find the one in particular, which will fulfill the conditions of the maximum sought. In the class of problems, moreover, to which the problem belongs, the function which is required to be a maximum or minimum, is not, as in the case of the ordinary problems, composed solely of finite quantities, but, in general, is the integral merely indicated of a differential expression not susceptible of integration.

173. We now proceed to explain the nature of the new auxiliary quantities required in the solution of the species of problems under consideration. In order to this, let it be supposed that the curve AMND, in the preceding figure, undergoes a transformation infinitely small so as to become AM′ N′D. Each point M′ of the new curve may be regarded as a point M of the other, which, in the process of the transformation, has passed from M to M′; so that each point of the new curve has its corresponding point in the other.

In this transformation each of the quantities which belongs to the system, undergoes a change in passing from the first to the second state which is called its *variation*. Thus, the variation of PM is P′M′ — PM, the variation of QN is Q′N′ — QN, that of the entire curve is AM′N′R′D — AMNRD, and so on.

The variation of a quantity, although the difference of two values nfinitely near each other of the same quantity, is not to be confounded with its differential. For the latter is the difference between two consecutive values of the same quantity taken upon the *same* curve; while the former is the difference of two values taken *from one curve to another.* Thus the differential of the ordinate PM is QN — PM, while its variation is P′M′ — PM.

To distinguish the variations from the ordinary differentials we employ the characteristic δ. Thus, the variation of x is indicated by δx, that of y by δy, and so on.

Derivation of the Variations.

174. In order to employ the new auxiliary quantities, we must, as in the case of the ordinary differentials, determine the manner in which they are derived from the primitive quantities, for which they are substituted in forming the equation of a problem.

From what has been done it will be perceived that the variations are derived from their primitives precisely in the same manner as the ordinary differentials. Thus, to find the variation of x, we consider x as becoming $x + \delta x$; and subtracting the first value of the quantity from the second we have δx for the variation sought. That is, to find the variation of a simple variable, we merely place the letter δ before it.

In like manner to find the variation of x^2, regarding x as becoming $x + \delta x$, we have for the second value of the quantity

$$(x + \delta x)(x + \delta x) = x^2 + 2x\delta x + \delta x^2.$$

$$\text{Subtracting the first } = x^2$$

$$\text{Variation} \quad = \quad 2x\delta x + \delta x^2.$$

But δx^2 being an infinitely small quantity of the second order, must be omitted by the side of $2x\delta x$; and we have, therefore, $\delta x^2 = 2x\delta x$.

In general to find the variation of a function we find its differential by the common rules, and then replace the letter d by δ. The variation of a quantity is, therefore, merely its differential taken in a new point of view. We pass to some examples.

Ex. 1. To find the variation of xy. Ans. $x\delta y + y\delta x$.

Ex. 2. To find the variation of $\frac{dy}{dx}$. Ans. $\frac{dx\delta dy - dy\delta dx}{dx^2}$.

Ex. 3. To find the variation of $(dx^2 + dy^2)^{\frac{1}{2}}$.

Ans. $\dfrac{dx\delta dx + dy\delta dy}{(dx^2 + dy^2)^{\frac{1}{2}}}$.

175. We thus see the manner in which the variations are derived. Between these and the differentials there are certain important relations which we will now explain.

If the figure be examined with attention, it will be observed that we may pass from the point M to N′ in two different ways. 1°. By proceeding from M to M′ by the method of variations, and then from M′ to N′ by the differential of the curve AM′N′. 2°. From M to N by the differential of the curve AMN, and then from N to N′ by the method of variations.

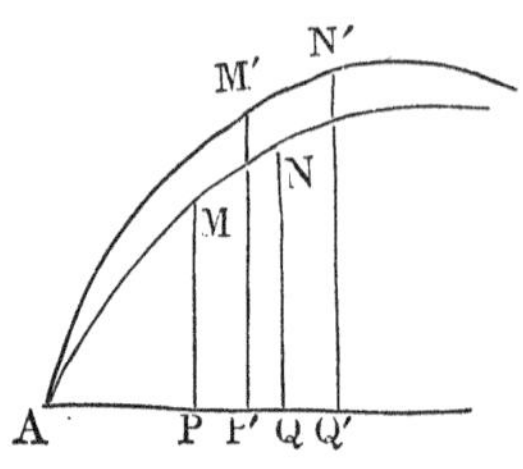

Let AP $= x$, PQ $= dx$, PP′ $= \delta x$; then P′Q′ $= d$AP′ along the curve AMN; whence P′Q′ $= dx + d\delta x$, and

$$AQ' = AP' + P'Q' = x + \delta x + dx + d\delta x. \quad (1)$$

Again, $$QQ' = \delta AQ = \delta x + \delta dx; \text{ and}$$

$$AQ' = AQ + QQ' = x + dx + \delta x + \delta dx. \quad (2)$$

Equating these two expressions for the same thing, and reducing, we obtain $$\delta dx = d\delta x.$$

In a manner altogether similar we find

$$\delta dy = d\delta y.$$

That is, *the variation of the differential of any quantity is equal to the differential of its variation.*

From this principle we also derive another analogous to it, viz. $\delta \int P = \int \delta P$; P denoting any function whatever of the variables x, y, z, &c., and their differentials.

Indeed, let $\int P = U$; then by differentiation $P = dU$; and taking the variation $\delta P = \delta dU$, or, by the preceding principle δP

$= d\delta U$. Taking next the integral of this last $\int \delta P = \delta U$; or substituting for U its value,

$$\int \delta P = \delta \int P.$$

That is, *the variation of the integral of any differential quantity whatever, is equal to the integral of its variation.*

These are fundamental principles in the method of variations.

Solution of Problems by aid of Variations.

176. We proceed next to show the aid derived from the auxiliaries now obtained, in the solution of some problems of maxima and minima of the kind to which we have just referred.

In the solution of such problems, it will be observed, we have the same general principle as in the ordinary problems of maxima and minima. Thus, let $\int y dx$ be the function which is required to be a maximum or minimum; then $\delta \int y dx = 0$; for the function having reached its maximum or minimum is incapable of further increase or diminution, and its variation must, therefore, be 0.

PROBLEM I.

177. It is required to find the curve of *quickest descent*, or that along which a heavy body will descend from one point to another in the shortest time possible.

Let M, m' be the two points; the question is to find among all the curves which can be described between M and m', that along which the body will descend in the least possible time.

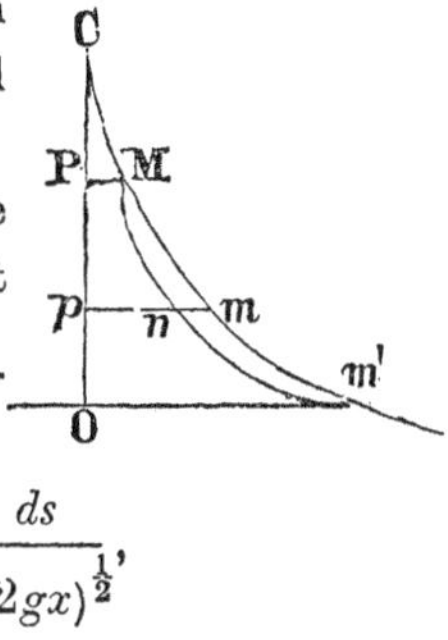

Let $CP = x$, $PM = y$, $CM = s$. The velocity of the body moving in the curve at M will be $(2gx)^{\frac{1}{2}}$, g being the force of gravity; whence

$$\frac{ds}{dt} = (2gx)^{\frac{1}{2}}, \text{ or } dt = \frac{ds}{(2gx)^{\frac{1}{2}}},$$

which, Mm being taken equal to ds, will be the time employed in moving from M to m.

Let $Cp = x + dx = x'$, $pm = y + dy = y'$, and $Cm = s + ds = s'$; then mm' being equal to ds', the time employed in moving from m to m', will be $\frac{ds'}{(2gx')^{\frac{1}{2}}}$; and the whole time of moving from M to m' will be

$$\frac{ds}{(2gx)^{\frac{1}{2}}} + \frac{ds'}{(2gx')^{\frac{1}{2}}}.$$

But, by the question, this should be a minimum; whence, omitting the constant factor,

$$\delta \frac{ds}{x^{\frac{1}{2}}} + \delta \frac{ds'}{x'^{\frac{1}{2}}} = 0. \qquad (1)$$

Substituting for ds, ds', their values in terms of dx, dy, and dx', dy'; taking the variation in reference to y only, since x and x' are the same for all curves that can be drawn between M and m'; and then restoring ds, ds' in the denominators, we obtain

$$\frac{dy}{x^{\frac{1}{2}}ds}\,\delta dy + \frac{dy'}{x'^{\frac{1}{2}}ds'}\,\delta dy' = 0. \qquad (2)$$

Again, whatever the curves may be, the ordinate Om' is the same for all. Thus we shall have $dy + dy'$ constant, and by consequence $\delta(dy + dy') = 0$; hence $\delta dy = -\delta dy'$. The equation (2), therefore, by this means reduces to

$$\frac{dy}{x^{\frac{1}{2}}ds} - \frac{dy'}{x'^{\frac{1}{2}}ds'} = 0;$$

or

$$\frac{dy}{x^{\frac{1}{2}}ds} = \frac{dy'}{x'^{\frac{1}{2}}ds'}. \qquad (3)$$

We have, therefore, $\frac{dy}{x^{\frac{1}{2}}ds}$ the same for every point of the curve, or equal to a constant. Let A be this constant; then

$$\frac{dy}{x^{\frac{1}{2}}ds} = \text{A}. \qquad (4)$$

To determine A it will be observed that $\frac{dy}{ds}$ expresses the sine of the angle which the tangent line makes with the axis of abscissas. Let a be the value of x at the point where the tangent is horizontal; at this point the angle will be a right angle and $\frac{dy}{ds} = 1$; hence

$$A = \frac{1}{a^{\frac{1}{2}}}; \text{ and } \frac{dy}{ds} = \left(\frac{x}{a}\right)^{\frac{1}{2}}; \quad (5)$$

But $ds = (dx^2 + dy^2)^{\frac{1}{2}}$; whence by substitution in (5)

$$\frac{dy}{dx} = \frac{x^{\frac{1}{2}}}{(a - x)^{\frac{1}{2}}}. \quad (6)$$

Integrating this last, putting $a = 2r$, we obtain

$$y = \text{ver-sin}^{-1}x - (2rx - x^2)^{\frac{1}{2}}, \quad (7)$$

which, art. 142, is, in finite terms, the equation to the Cycloid.

We have, therefore, the Cycloid for the curve of quickest descent, or the *Brachystochrone.*

Reviewing the course pursued, it is obvious that

1°. By aid of the new auxiliaries, we have been able to find, with great facility, an equation for the problem, viz. the equation (1).

2°. By the conditions of the problem, we have next been able to eliminate these auxiliaries, after they have served their purpose; and have thus, equation (5), obtained a result in terms of the variables and their differentials only.

3°. The solution being thus reduced to the ordinary Calculus, we next eliminate, in the usual way, the differentials found in this result, and obtain finally an equation for the curve sought in terms of the variables only and the constants which enter into it.

So far as the mere solution of the present problem is concerned, we might have stopped at equation (6), since we had previously found this to be, art. 142, the differential equation of the cycloid. The course pursued exhibits in connection the manner in which both the species of auxiliaries necessary to the solution of the question are eliminated, and the result, freed entirely from the indeterminates employed, left in finite and determinate terms.

PROBLEM II.

178. Let it be proposed, as a second problem, to find among all the plane curves which can be drawn between two points that which is the shortest.

Let x', y' and x'', y'' be the co-ordinates of the two points, and s the curve sought; then $ds = (dx^2 + dy^2)^{\frac{1}{2}}$, and we have by the question

$$\delta \int (dx^2 + dy^2)^{\frac{1}{2}} = 0, \text{ or } \int \delta\, (dx^2 + dy^2)^{\frac{1}{2}} = 0. \quad (1)$$

Taking the variation, transposing the δ, and restoring ds in the denominator, we obtain

$$\int \frac{dx}{ds}\, d\delta x + \int \frac{dy}{ds}\, d\delta y = 0. \quad (2)$$

Integrating next by parts, we find

$$\frac{dx}{ds}\, \delta x + \frac{dy}{ds}\, \delta y - \int \left(d\,\frac{dx}{ds}\, \delta x + d\,\frac{dy}{ds}\, \delta y \right) = 0. \quad (3)$$

This equation consists, it is evident, of two distinct parts, one free from the sign of integration, and the other subject to it. The latter not being integrable so long as δx, δy preserve the independence which the nature of the question requires, its terms in x and y cannot operate the destruction of those in x and y of the other part. In order, therefore, that equation (3) may be satisfied, its two parts must be separately equal to 0.

Considering first the part of (3) under the sign of integration, and removing this sign, since if a function is equal to 0 its differential will also be equal to 0, we shall have

$$d\,\frac{dx}{ds}\, \delta x + d\,\frac{dy}{ds}\, \delta y = 0; \quad (4)$$

but, δx, δy being independent, in order that this equation may be satisfied its two terms must also be separately equal to 0, and we shall have

$$d\frac{dx}{ds}=0, \quad d\frac{dy}{ds}=0.$$

From the first of these we obtain $\frac{dx}{ds}=C$. Substituting in this for ds its value, $(dx^2+dy^2)^{\frac{1}{2}}$, and putting C' for the constant, it becomes $\frac{dy}{dx}=C'$. From which we obtain by integration

$$y=C'x+C'',$$

which is the equation to a straight line.

Returning next to the part of equation (3) without the sign of integration, we shall have for this part, the integral being taken between the two points $x', y', \; x'', y''$,

$$\frac{dx''}{ds''}\delta x''+\frac{dy''}{ds''}\delta y''-\frac{dx'}{ds'}\delta x'-\frac{dy'}{ds'}\delta y'=0, \qquad (5)$$

an equation which contains only the values of $\delta x, \delta y$ which pertain to the limits of the integration. Thus the two parts of equation (3) differ essentially in their signification; the part under the sign of integration establishing a general relation between the variables x and y, while the part without connects the particular values which these quantities have at the limits of integration.

The proposed problem admits of two cases, 1°, when the two points between which the line of shortest distance is to be drawn are fixed; 2°, when the two points are situated each upon a known curve.

In the first case, the variations of the co-ordinates x', y', x'', y'' being each equal to 0, the terms of equation (5), or which is the same thing those of (3) without the sign of integration, vanish of themselves. Thus the equation (4) is sufficient for the complete solution of the problem. All that is necessary is to draw between the two given points a *straight line*, which by this equation is determined to be the shortest distance between them.

II. For the second case the problem may be enunciated as follows.

Two plane curves being given by their equations, to determine the shortest distance between them.

From the preceding investigation the curve of shortest distance is already found to be a straight line. It only remains to determine the points on the curves through which the line must be drawn. The co-ordinates of the points, x', y', x'', y'', which are unknown, must, it is evident, be subject to the two following conditions; 1°. they must, with their variations, satisfy the first part of equation (3) put equal to 0, the integral being taken, equation (5), between the two points as limits; 2°. they must satisfy the equations to the given curves.

Let $dy = mdx$, $dy = ndx$ be the differential equations, respectively, of the curves upon which the two points are situated. Then, by the second condition, since the variations will have necessarily the same relations as the differentials of the curves, we shall have,

$$\delta y' = m\delta x', \quad \delta y'' = n\delta x''; \qquad (6)$$

whence by substitution in (5)

$$\left(\frac{dx''}{ds''} + n\frac{dy''}{ds''}\right)\delta x'' - \left(\frac{dx'}{ds'} + m\frac{dy'}{ds'}\right)\delta x' = 0.$$

But the variations $\delta x''$, $\delta x'$ being independent, the two parts of this equation must be separately equal to 0, and we have

$$dx'' + ndy'' = 0, \quad \text{or } \frac{dy''}{dx''} = -\frac{1}{n}, \qquad (7)$$

$$dx' + mdy' = 0, \quad \text{or } \frac{dy'}{dx'} = -\frac{1}{m}; \qquad (8)$$

whence by substitution from (6)

$$\frac{\delta y''}{\delta x''}\frac{dy''}{dx''} = -1, \quad \frac{\delta y'}{\delta x'}\frac{dy'}{dx'} = -1.$$

From which it follows that the right line already determined to be the shortest that can be drawn, *must make a right angle with each of the two given curves.*

From the equation $y = C'x + C''$ we have $dy = C'dx$. And since the co-ordinates x', y', x'', y'' must satisfy this equation, we obtain by means of (7) and (8),

$$1 + C'm = 0, \qquad 1 + C'n = 0. \qquad (9)$$

But a right line passing through the points x', y', x'', y'', has for its equation

$$y - y' = \frac{y'' - y'}{x'' - x'}(x - x');$$

in this case, therefore, $C' = \dfrac{y'' - y'}{x'' - x'}$; whence by substitution in (9)

$$x'' - x' + m(y'' - y') = 0, \qquad x'' - x' + n(y'' - y') = 0. \quad (10)$$

Combining next these equations with the primitive equations to the given curves, the unknown quantities x', y', x'', y'' will be determined. We thus find the points on the curves, through which if a straight line be drawn it will be the shortest distance between them. The problem is thus completely solved.

Ex. 1. Let $y = x + 3$, $y^2 = x$ be two lines referred to the same axes; to find on these lines the points, through which if a right line be drawn it will be the shortest distance between them.

Let the co-ordinates of the point on the first line be x', y', those of the point on the second x'', y''; the equations to the lines will then be $y' = x' + 3$, $y''^2 = x''$, from which we obtain

$$n = \frac{dy''}{dx''} = \frac{1}{2y''}; \quad m = \frac{dy'}{dx'} = 1;$$

whence by substitution and elimination in (10) we find $x', y' = -1.125, 1.875$; $x'', y'' = 0.25, 0.5$, the co-ordinates of the points required.

Ex. 2. Two bodies are moving, one on the line $y' = x + 1$, the other on the curve $x''^2 + (y'' - 3)^2 = 1$. At what points must they be, at the same time, in order to approach each other the nearest possible?

179. The preceding problems belong to a class denominated maxima and minima *absolute*. We proceed next to a problem of a different class called maxima and minima *relative*.

PROBLEM III.

To find the nature of the curve, of a given length, which shall inclose the greatest area.

Let s be the length of the curve sought, then

$$\int (dx^2 + dy^2)^{\frac{1}{2}} = s;$$

and by the first condition

$$\delta \int (dx^2 + dy^2)^{\frac{1}{2}} = 0. \quad (1)$$

Again, for the area of the curve we have $\int y dx$; whence by the second condition

$$\delta \int y dx = 0. \quad (2)$$

Multiplying equation (1) by an arbitrary constant n, and adding it to (2), we have

$$\delta \int (y dx + n (dx^2 + dy^2)^{\frac{1}{2}}) = 0, \quad (3)$$

an equation which implies that (1) and (2) are necessarily true, and which is sufficient for the solution of the problem.

The process will now be the same as in the preceding problem. Performing the operations, we have for the part under the sign of integration

$$-\int \left(\left(dy + nd\frac{dx}{ds}\right)\delta x - \left(dx - nd\frac{dy}{ds}\right)\delta y\right);$$

which gives to determine the curve sought

$$dx - nd\frac{dy}{ds} = 0;$$

whence, by integration, $\quad x - n\frac{dy}{ds} = C;$

which, substituting for ds its value $(dx^2 + dy^2)^{\frac{1}{2}}$, may be put under the form

$$dy = \frac{(x - C)\,dx}{(n^2 - (x - C)^2)^{\frac{1}{2}}};$$

or, taking the integral

$$y = -(n^2 - (x - C)^2)^{\frac{1}{2}},$$

or

$$n^2 = y^2 + (x - C)^2;$$

which is the equation to the circle. Thus the curve sought will be the circle.

180. Problems of the kind we have solved by the aid of variations have received the name of *Isoperimetrical Problems;* though the designation applies more especially to those of the class last considered. They engaged much the attention of geometers about the time of the invention of the Calculus. The results obtained led, however, to no general rules, until La Grange conceiving, from a consideration of them, the idea of the method of variations, reduced their solution to a common and general method.

The method of variations, though deriving its origin from this class of problems, is not confined to them. Indeed, it is, from its nature, manifestly adapted to all questions in which the magnitudes to be considered may be made to vary in two different ways, or which require in their solution the use of differentials taken in two distinct and different points of view. It is from this consideration that the method is peculiarly adapted to the problems of rational mechanics and physical astronomy.

The method, it will be observed, is not to be regarded as a new Calculus, but rather as an extension of the ordinary Calculus. By means of the auxiliaries introduced by the latter, we are enabled to solve, with facility, problems the solution of which entirely transcends the powers of common algebra. By aid of the auxiliaries provided by the method of variations, we again bring within our grasp problems which, in their turn, baffle the powers of the ordinary Calculus. In the two we have a logical instrument the most powerful the human mind has yet constructed.

SECTION XVIII.

APPLICATION TO ASTRONOMY.

181. In what precedes, we have seen the application of the Calculus to various problems relating to force and motion. We shall now apply it to the two following problems of Astronomy.

1°. The heavenly bodies being considered as spheres, to determine their mutual attraction.

2°. To determine the law of the force which confines the planets and comets to their orbits.

Before proceeding to the first of these questions, we take the following more simple problems of attraction.

PROB. I. To find the attraction of a uniform circular arc on a particle in the centre of the circle, the force varying in the inverse ratio of the square of the distance.

Let C be the centre, BAD the arc, A its middle point. Join CA, this will be the line in which the resultant of the attraction acts. Let Mm be any infinitely small portion or element of the arc. Put the angle BCA $= \omega$, MCA $= \theta$, AC $= a$. Then MC$m = d\theta$; and the mass of a unit of length of the arc being regarded as unity, the mass of the arc Mm will be expressed by $a d\theta$. The attractive force of Mm on C, in the direction MC, will then be

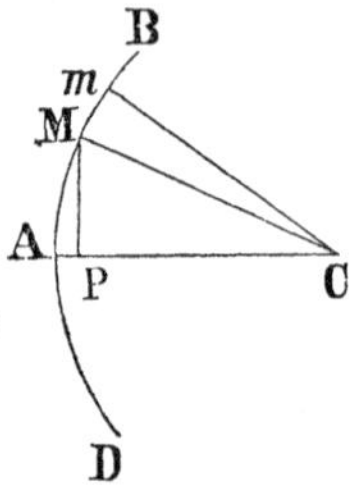

$$\frac{a\,d\theta}{a^2} = \frac{d\theta}{a}.$$

Draw PM perpendicular to AC; then, to find the attractive force of Mm in the direction AC, we must multiply the force in the direction CM by $\frac{\text{CP}}{\text{CM}}$ or $\cos\theta$. Putting A for the whole attraction, we shall have for its element

$$d\text{A} = \frac{\cos\theta\, d\theta}{a}.$$

Integrating, and taking the integral between the limits $\theta = \omega$, $\theta = -\omega$, we have for the attraction of the whole arc

$$\frac{2\sin\omega}{a}.$$

PROB. II. To find the attraction of a circle on a particle situated in a line perpendicular to its plane, and passing through its centre.

Let S be the centre of the circle; P the attracted point; $SP = r$, $Pm = f$; Sm any radius $= x$, and SQ any radius indefinitely nearly equal to this, so that $mQ = dx$. Let the angle $ASQ = \omega$; and, the radius SR being drawn infinitely near SQ, the angle $QSR = d\omega$; then the quadrilateral $QmnR = xd\omega dx$.

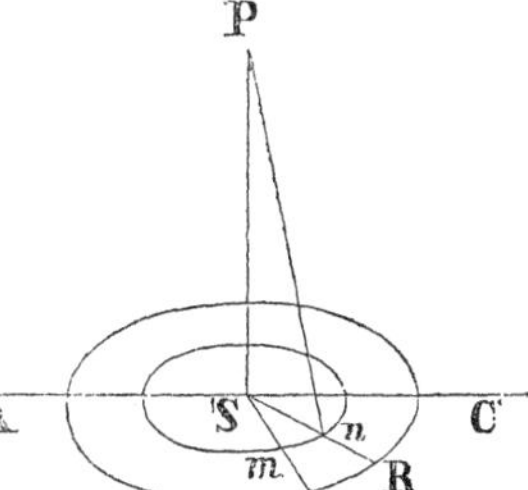

Let the law of attraction be represented by $\phi(f)$; then the attraction of $QmnR$ will be expressed by $xd\omega dx\,\phi(f)$. And this being resolved in the direction PS, we have for the element dA of the attraction,

$$dA = xd\omega dx\,\phi(f)\,\frac{r}{f};$$

whence $$A = \int\int xd\omega dx\,\phi(f)\,\frac{r}{f};$$

in which $f = (r^2 + x^2)^{\frac{1}{2}}$. Let $\phi(f) = f^n$; then, taking the integral from $\omega = 0$ to $\omega = 2\pi$, we obtain

$$A = \int 2\pi r x dx\,(r^2 + x^2)^{\frac{n-1}{2}}.$$

Integrating next from $x = 0$ to $x = a$, and determining the constant on the hypothesis $x = 0$, we have

$$A = \frac{2\pi r}{n+1}[(r^2 + a^2)^{\frac{n+1}{2}} - r^{n+1}].$$

Ex. Let $n = -2$, or the force vary inversely as the square of the distance. Then

$$A = 2\pi\left(1 - \frac{r}{(r^2 + a^2)^{\frac{1}{2}}}\right).$$

Attraction of Spheres.

182. Returning now to our object, let the following problem be proposed.

A spherical shell of inconsiderable thickness is composed of particles which attract in the inverse ratio of the square of the distance; to find the attraction upon a particle without the shell.

Let S be the centre of the spherical shell, So its interior radius $= y$, Sn its exterior radius $= y + dy$. Suppose two planes AGC, AnC, passing through AC, to make with each other an infinitely small angle G$pn = d\omega$, the plane of Gpn being perpendicular to AC. Let the angle PSG $= \theta$, and GSB $= d\theta$. Then the arc G$n = y \sin \theta d\omega$, and the arc BG $= y d\theta$. We have, therefore, for the content of the infinitely small solid Bo,

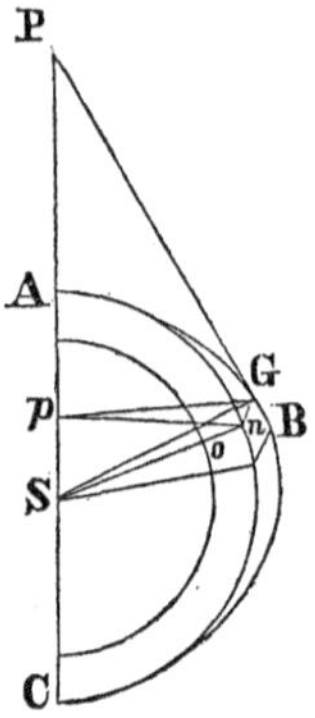

$$y^2 \sin \theta \, d\theta \, d\varpi \, dy.$$

Let SP $= b$, PG $= f$. Then the solid Bo being infinitely small, its attraction may be regarded as that of a single particle; hence we shall have for the element of attraction, dA, in the direction PG,

$$dA = y^2 dy \sin \theta \, d\theta d\varpi \times \frac{1}{f^2},$$

$$= - y^2 dy \, d\omega d \cos \theta \times \frac{1}{f^2}.$$

But $f^2 = b^2 - 2by \cos \theta + y^2$. Differentiating this equation in relation to θ, θ and y being independent, we have

$$fdf = - byd \cos \theta,$$

and by substitution

$$dA = \frac{ydyd\omega}{b} \cdot \frac{df}{f}.$$

Reducing this to the direction PS, since the attraction manifestly acts in this direction, we multiply it by cos SPG, equal, Trig. art. 81, to $\frac{b^2 + f^2 - y^2}{2bf}$; whence

$$dA = \frac{ydyd\omega df}{2b^2} \cdot \frac{b^2 + f^2 - y^2}{f^2};$$

and $$A = \iiint \frac{y\,dy\,d\omega}{2b^2} \cdot \frac{(b^2 + f^2 - y^2)\,df}{f^2}.$$

Integrating first from $\varpi = 0$ to $\varpi = 2\pi$, we have

$$A = \iint \frac{\pi y\,dy}{b^2} \cdot \frac{(b^2 + f^2 - y^2)\,df}{f^2}.$$

Integrating next from $f = b - y$ to $f = b + y$, we obtain

$$A = \int \frac{4\pi}{b^2} y^2 dy.$$

Designating the internal and external radii of the shell, So and Sn, by r, r' respectively, and integrating finally between the limits r and r', we have for the whole attraction of the shell

$$A = \frac{4\pi}{3b^2}(r'^3 - r^3).$$

But the mass of the shell is equal to the difference of the masses of two spheres, whose radii are r and r' respectively; hence, putting M for this mass,

$$M = \frac{4\pi}{3}(r'^3 - r^3),$$

and we shall have, therefore,

$$A = \frac{M}{b^2}.$$

Thus the attraction is equal to the mass of the shell divided by the square of the distance from the centre; and is, therefore, the same as if the whole mass were concentrated in a single particle at the centre.

If we suppose $r = 0$, we obtain the attraction of a sphere whose radius is r'. Thus, the density of the sphere being uniform, the result is the same for the whole sphere as for the shell.

If the density varies according to any law, we may consider the sphere as composed of concentric shells with densities varying according to the given law. The law of attraction above obtained being true for each of the shells separately, will be true also for the whole sphere which is composed of them.

The heavenly bodies are very nearly spheres. Considered as such they attract, therefore, *as if their whole masses were concentrated in a single particle at their centres.*

Motions of the Celestial Bodies. Law of the force.

183. The laws which regulate the motions of the planets about the sun, are known under the denomination of the laws of Kepler; having been discovered by that distinguished Astronomer, who deduced them from innumerable observations. They are three in number and are as follows.

1°. The planets and comets move in plane curves, and their radii vectors describe about the sun, as a centre, areas proportional to the times.

2°. The orbits, or trajectories of the planets and comets are conic sections, having the sun in one of their foci.

3°. The squares of the times of the revolutions of the planets about the sun, are proportional to the cubes of their mean distances from the sun.

184. From these laws, derived from observation, Newton deduced the law of universal gravitation. We shall give an elementary view of the aid derived from the Calculus in the accomplishment of this great work.

Let AmP be the elliptical orbit of a planet m, having the centre of the sun in the focus S. Let SX, SY be two rectangular axes in the plane of the planet's orbit, and having their origin at S. Draw Sm, and let fall mp perpendicular upon SX. Sm which joins the centres of the sun and planet is the *radius vector;* and Sp, pm equal to x and y respectively, will be the co-ordinates of the planet at any point m of the orbit.

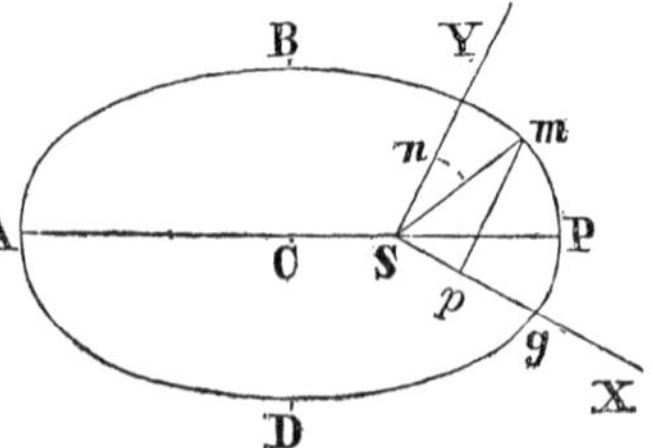

Let X and Y be the components, in the directions of AX and AY respectively, of the force F which acts upon the planet. Then we shall have for the equations of the planet's motion,

$$\frac{d^2x}{dt^2}=X, \quad \frac{d^2y}{dt^2}=Y. \qquad (1)$$

If we now multiply the first of these two equations by $-y$, and the second by x, and add them, we obtain

$$\frac{xd^2y-yd^2x}{dt^2}=Yx-Xy,$$

or

$$\frac{d(xdy-ydx)}{dt^2}=Yx-Xy. \quad (2)$$

Let A be the area of any sector PSg of the ellipse. Supposing a perpendicular Pn let fall from P on the axis SY, the area of the quadrilateral PnSg, thus formed, will be comprised between the limits $y=0$, $y=\text{S}n$, and we shall have area PnS$g=\int xdy$. But the area of the triangle PnS $=\frac{1}{2}xy$; hence

$$A=\int xdy-\tfrac{1}{2}xy,$$

and

$$dA=xdy-\tfrac{1}{2}xdy-\tfrac{1}{2}ydx$$
$$=\tfrac{1}{2}(xdy-ydx.)$$

Let $c=$ twice the area described by the radius vector in the unit of time; then by the first law of Kepler

$$xdy-ydx=cdt; \qquad (3)$$

whence

$$\frac{d(xdy-ydx)}{dt^2}=0;$$

and by substitution in (2) $Yx-Xy=0;$ (4)

whence

$$X:Y::x:y::\text{S}p:pm.$$

The forces X and Y being thus proportional to the co-ordinates Sp, pm, the resultant of these forces must pass through their origin, or the centre of the sun S. The curve described by the planet is concave toward the sun; the force, therefore, that causes the planet to describe that curve tends toward the sun.

The first deduction from the laws of Kepler is, then, the important one, *that the force which retains the planets in their orbits is directed toward the centre of the sun.*

185. We proceed next to determine the law of the force, or the manner in which it varies at different distances from the sun.

If the first of the equations (1) be multiplied by dx and the second by dy, we obtain for the sum

$$\frac{dxd^2x + dyd^2y}{dt^2} = \mathrm{X}dx + \mathrm{Y}dy.$$

The left hand member of this being a perfect differential, we have by integration

$$\frac{dx^2 + dy^2}{dt^2} = 2\int(\mathrm{X}dx + \mathrm{Y}dy).$$

Deducing the value of dt from (3) which expresses the law of the areas, and substituting, we have

$$\frac{c^2(dx^2 + dy^2)}{(xdy - ydx)^2} = 2\int(\mathrm{X}dx + \mathrm{Y}dy). \quad (5)$$

We next, for greater simplicity, transform this equation into a polar equation, having for its co-ordinates the radius vector Sm, and the polar angle mSg. Let S$m = r$, mS$g = v$; then

$$\mathrm{S}p = x = r\cos v,\ pm = y = r\sin v,\ r = (x^2 + y^2)^{\frac{1}{2}};$$

whence $dx^2 + dy^2 = r^2dv^2 + dr^2$. Also from the element of the area of polar curves, art. 127, and equation (3), we have $xdy - ydx = r^2dv$.

Again, let F be the resultant of the forces X and Y; then

$$\mathrm{F} = (\mathrm{X}^2 + \mathrm{Y}^2)^{\frac{1}{2}},\ \mathrm{X} = -\mathrm{F}\cos v, \text{ and } \mathrm{Y} = -\mathrm{F}\sin v;$$

the sign is negative, because the force F in the direction mS tends to diminish the co-ordinates. Multiplying the two last equations by dx and dy respectively, and adding, we have

$$\mathrm{X}dx + \mathrm{Y}dy = -\mathrm{F}(\cos v dx + \sin v dy);$$

from which by means of the equations $x = r\cos v$, $y = r\sin v$, we get

$$Xdx + Ydy = -Fdr;$$

and the equation (5) becomes, therefore, by substitution

$$\frac{c^2(r^2dv^2 + dr^2)}{r^4dv^2} + 2\int Fdr = 0; \qquad (6)$$

whence

$$dv = \frac{cdr}{r(-c^2 - 2r^2\int Fdr)^{\frac{1}{2}}}.$$

If the value of F were now known in terms of the distance r, this equation would give the nature of the curve, or trajectory, described by the planet. But it is the force F which is unknown that we wish to determine, the nature of the orbit being given. Differentiating, therefore, (6) from which the equation for dv is derived, we have, to determine F,

$$F = \frac{c^2}{r^3} - \frac{c^2}{2}\,\frac{d\left(\frac{dr^2}{r^4dv^2}\right)}{dr}. \qquad (7)$$

We have thus obtained an expression for the force F in terms of the radius vector Sm, and the corresponding variable angle mSg $= v$. But in order to determine the law of the force we must eliminate v, so that F will be expressed in terms of the radius vector r or Sm only.

By the second law of Kepler the planets and comets move in conic sections, having the sun in one of their foci. Let us employ, therefore, for the purpose of the required elimination the polar equation of the ellipse, viz.

$$r = \frac{a(1 - e^2)}{1 + e(\cos v - \varpi)}; \qquad (8)$$

in which e represents the eccentricity, $2a$ the greater axis, ϖ the angle PSg which this axis makes with the axis AX, and v the angle mSg as above.

This equation becomes that of a parabola, when $e = 1$ and a is infinite; and that of an hyperbola when e is greater than unity, and a is negative.

But this equation may be written

$$\frac{1}{r} = \frac{1 + e\cos(v - \varpi)}{a(1 - e^2)}. \qquad (9)$$

Taking the differential, dividing by dv, squaring both sides, and substituting for $\sin^2(v - \varpi)$ its equal $1 - \cos^2(v - \varpi)$, we obtain

$$\frac{dr^2}{r^4 dv^2} = \frac{e^2}{a^2(1-e^2)^2} - \frac{e^2\cos^2(v - \varpi)}{a^2(1 - e^2)^2}. \qquad (10)$$

But equation (9) gives

$$\frac{e^2\cos^2(v - \varpi)}{a^2(1 - e^2)^2} = \frac{1}{r^2} - \frac{1}{a^2(1 - e^2)^2} - \frac{2e\cos(v - \varpi)}{a^2(1 - e^2)^2},$$

or by substituting for $e\cos(v - \varpi)$ from (8)

$$\frac{e^2\cos^2(v - \varpi)}{a^2(1 - e^2)^2} = \frac{1}{r^2} - \frac{2}{ar(1 - e^2)} + \frac{1}{a^2(1 - e^2)^2}; \qquad (11)$$

whence by substitution in (10)

$$\frac{dr^2}{r^4 dv^2} = \frac{2}{ar(1 - e^2)} - \frac{1}{r^2} - \frac{1}{a^2(1 - e^2)}; \qquad (12)$$

and this being substituted in (7) gives, after performing the differentiation and reducing,

$$\text{F} = \frac{c^2}{a(1 - e^2)} \cdot \frac{1}{r^2}. \qquad (13)$$

But the coefficient $\frac{c^2}{a(1 - e^2)}$ being constant, the force, it is evident, will vary as $\frac{1}{r^2}$, or inversely as the square of the distance of the planet from the sun.

Thus, *the force which retains the planets and comets in their orbits is directed toward the sun, and varies in the inverse ratio of the square of the distance from the sun.*

186. The law thus deduced is obtained for each of the planets separately. We shall now show that the force is the same for all the planets at whatever distances they are placed from the sun.

The intensity of the force depends, it is evident, upon the coefficient $\frac{c^2}{a(1-e^2)}$, which may be found by Kepler's laws.

Let T denote the periodic time, or time of revolution of a planet. Then c denoting, as above, the double area described by the radius vector in the unit of time, the area described in the time T will be cT. But during the time T the radius vector will describe the whole ellipse, the area of which is $\pi a^2(1-e^2)^{\frac{1}{2}}$, a and $(1-e^2)^{\frac{1}{2}}$ denoting the semi-axes. Thus we shall have

$$c = \frac{2\pi a^2(1-e^2)^{\frac{1}{2}}}{\mathrm{T}}.$$

Putting $k = \frac{c^2}{a(1-e)^2}$, and substituting the value of c, we get

$$k = \frac{4\pi^2 a^3}{\mathrm{T}^2}.$$

In like manner if we consider another planet, we have

$$k' = \frac{4\pi^2 a'^3}{\mathrm{T}'^2}.$$

But by the third of Kepler's laws

$$\mathrm{T}^2 : \mathrm{T}'^2 :: a^3 : a'^3;$$

hence

$$k = k';$$

that is, the accelerating force is the same at the unit of distance for all the planets, and varies only from one body to another by reason of the distance. It follows, therefore, that if all the planets were placed at equal distances from the sun, they would, if left solely to the action of this force, fall in the same time to the sun; that which is analogous to what is observed in respect to bodies acted upon by attraction near the surface of the earth.

Proceeding with the investigation, it is proved that the force which confines the planets and comets to their orbits, is the same as that which causes the descent of a heavy body toward the surface of the earth. We thus reach the law of universal gravitation, viz.

that all the particles of matter attract each other as the masses directly, and as the squares of the distances inversely.

187. The law of the force being obtained, as we have seen, from the laws of Kepler derived from observation, we may next invert the problem ; and, assuming the law of gravitation to be the law of nature, deduce from the equations of motion derived from this principle all the circumstances of the motions of the heavenly bodies, and then compare the results with observation. This is what has been done by La Place in his immortal work, the *Mèchanique Cèleste.* Setting out with this single principle, the revolutions and rotations of the planets modified by their mutual disturbances, the oscillations of the fluids at their surfaces, the various changes the system has undergone in ages past, those it will undergo in ages to come, are made the subject of direct calculation. The astonishing accuracy of the results obtained exhibit alike the truth and the power of the instrument by which the work is accomplished.

SECTION XIX.

LIMITS. DERIVED FUNCTIONS.

188. Thus far, in accordance with the method of Leibnitz, we have employed, as auxiliaries to the establishment of equations, the infinitely small increments or elements of which the magnitudes under consideration are supposed to be composed. Newton has based the Calculus upon a different conception. From his point of view the auxiliaries employed to facilitate the formation of equations are the *limits of the ratios* of the simultaneous increments of the primitive quantities which require to be considered; these limits, or which is the same thing, *final ratios* of the increments having always a finite and determinate value.

By the *limit* of a proposed magnitude we understand that to which the magnitude may be made to approach indefinitely near.

Thus the circle is the limit to the polygon inscribed; since by increasing the number of its sides the polygon may be made to approach indefinitely near to the circle, or to differ from it as little as we please.

A magnitude is said to be *ultimately equal* to its limit; and the two are said to be *ultimately in a ratio of equality.*

A line or figure *ultimately coincides* with the line or figure which is its limit.

189. To give an elementary idea of the Calculus according to the method of Newton, we resume the problem of tangents, viz. To determine for each point of a plane curve, the equation to which is given, the direction of its tangent.

Let AMM′ be the curve, M the point to which the tangent is required to be drawn. Let T′MM′ be a secant line cutting the curve in M and also in the point M′ near to M. Supposing the secant to revolve about the point M, as the point M′ approaches M, the secant T′MM′ will approach, it is evident, more nearly to the tangent TM; with which it will coincide when the point M′ falls upon M. The tangent line is, then, the *limit* to the secant. Let AP, AP′ and PM, P′M′ be the co-ordinates respectively of the points M and M′, and let the difference between the abscissas AP, AP′ be represented by Δx, and the difference between the ordinates PM, PM′, by Δy. Then, t being put for the trigonometrical tangent of the angle MT′P which the secant line makes with the axis of x, we shall have

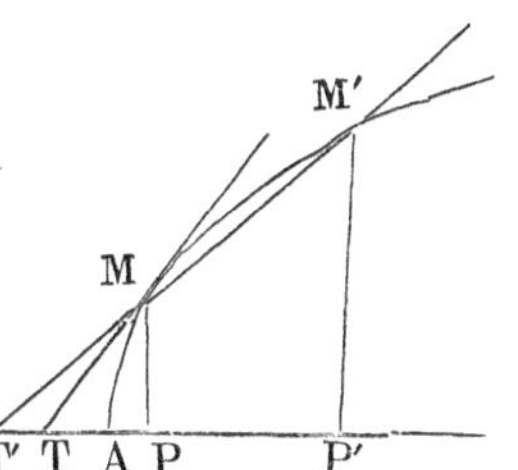

$$t = \frac{\Delta y}{\Delta x}.$$

Passing now to the limit, and denoting the limit by L, we have

$$t = \mathrm{L}\frac{\Delta y}{\Delta x},$$

for the trigonometrical tangent of the angle which the tangent sought makes with the axis of x. For the complete solution of the problem we have now only to eliminate the auxiliary employed; or, in other words, to find, for each particular case, the value of the limit $L\frac{\Delta y}{\Delta x}$; which is easily done when the equation to the curve is given.

Let, for example, the equation to the curve be $y = ax^2$. We shall then have

$$y + \Delta y = a(x + \Delta x)^2 = ax^2 + 2ax\Delta x + a\Delta x^2;$$

whence $$\frac{\Delta y}{\Delta x} = 2ax + a\Delta x.$$

Here, it is evident, that the limit toward which the second member of this equation tends, as Δx diminishes, is $2ax$. Putting, therefore, $\Delta x = 0$, in order to pass to the limit, we have $t = 2ax$. The direction of the tangent, is therefore, known.

190. The nature of the auxiliaries being now understood, we proceed to show the manner in which they are derived from their primitives.

In order to this, let $u = f(x)$; and let u' denote what u becomes when x has received an arbitrary increment h. Let $f(x+h)$ be developed according to the ascending powers of h; and let A, B, C, &c., which are functions of x, be the coefficients of these powers. Then, if we suppose $h = 0$, we have $A = u$, and the development may be written

$$u' = u + Bh + Ch^2 + Dh^3 + \&c.; \quad (1)$$

whence $$\frac{u' - u}{h} = B + Ch + Dh^2 + \&c.$$

The right hand member of this equation approaches, it is evident, more nearly to B as h is diminished; and becomes equal to B when $h = 0$. Passing, then, to the limit we have

$$L\frac{u' - u}{h} = B. \quad (2)$$

In general, therefore, to find the limit of the ratio of the increment of a function to that of its variable, *we develop the function according to the ascending powers of the increment of its variable, subtract its primitive value, divide the remainder by the increment of the variable, and take the first term of the quotient.*

The equation (1) may be put under the form

$$u' - u = Bh + (C + Dh + \&c.)h^2,$$

or putting P for B and P′ for the coefficient of h^2,

$$u' - u = Ph + P'h^2, \qquad (3)$$

a formula which is of frequent use.

191. The corresponding increments $u' - u$, and h, considered as indefinitely small, are called the *differentials* of u and x; and are written du, dx respectively. Employing this notation, the limit, equation (2), may be written

$$\frac{du}{dx} = B.$$

The value B of this limit is called the *differential coefficient of* u. The differential coefficient of a function multiplied by the differential of its variable is called the differential of the function. Thus, if B is the differential coefficient, Bdx is the differential of the function u. Let $u = ax^3$, for example; the differential coefficient will be $3ax^2$, and the differential will be $3ax^2dx$.

1. Proceeding now with our purpose, let it be required to find the differential coefficient of the function $u = y + z - v$, where y, z and v are functions of x.

Giving to x an increment h we have

$$u' - u = (y' - y) + (z' - z) - (v' - v),$$

or by the formula (3)

$$u' - u = (Ph + P'h^2) + (Qh + Q'h^2) - (Rh + R'h^2);$$

from which, dividing by h, and passing to the limit, we have

$$\frac{du}{dx} = P + Q - R.$$

Whence, *the differential coefficient of a function, composed of the sum or difference of several functions of the same variable, is equal to the sum or difference of the differential coefficients of these functions.*

2. Let u and v be two functions dependent upon the same variable x, to find the differential coefficient of their product.

By the formula (3) we have

$$u' = u + Ph + P'h^2; \quad v' = v + Qh + Q'h^2.$$

Multiplying these equations, member by member,

$$u'v' = uv + vPh + uQh + PQh^2 + \&c.;$$

whence, transposing, dividing by h, and passing to the limit,

$$\frac{d(uv)}{dx} = vP + uQ.$$

Hence, *the differential coefficient of the product of two functions of the same variable, is equal to the sum of the products of each of them by the differential coefficient of the other.*

3. Let $u = \frac{s}{v}$, where s and v are functions of the same variable x. We shall then have $uv = s$, and

$$\frac{d(uv)}{dx} = \frac{ds}{dx}; \text{ whence } \frac{ds}{dx} = v\frac{du}{dx} + u\frac{dv}{dx}.$$

Transposing, dividing by v, substituting for u its value, and reducing,

$$\frac{du}{dx} = \frac{v\frac{ds}{dx} - s\frac{dv}{dx}}{v^2}.$$

Whence, to find the differential coefficient of a fraction, whose numerator and denominator are functions of the same variable, *multiply the differential coefficient of the numerator by the denominator, and the differential coefficient of the denominator by the numerator; subtract the second product from the first and divide the remainder by the square of the denominator.*

4. Let $u = ax^n$, n being entire or fractional, positive or negative. Developing by the binomial theorem, transposing and passing to the limit, we obtain

$$\frac{du}{dx} = max^{n-1}.$$

To find, therefore, the differential coefficient of the power of a single variable, *multiply by the exponent and diminish the exponent by unity.*

192. The above rules are sufficient for all algebraic functions. In their application it will be recollected, as may easily be shown, that *the differential coefficient of* u + C, C *being any constant quantity and* u *being any function of* x, *is the same as the differential coefficient of* u; *and the differential coefficient of* Au *is equal to the differential coefficient of* u *multiplied by* A.

The process for finding the differential coefficient, or, in general, the differential, is called *differentiation*. The reverse process is called *integration*, and is indicated by the sign $\int$.

From what has been done, each rule for finding the differential coefficient, being inverted, will, it is evident, furnish a rule by which to find the function from the differential coefficient. Thus, for example, $\int ax^3 = \frac{1}{4}ax^4$.

193. We have seen how the problem of tangents is solved by the method of Newton. We pass to some additional examples.

Ex. 1. A plane curve being given by its equation, to find its area.

Let AMN be the curve, $AP = x$, $PM = y$, $PP' = h$. Then when x receives an increment h, the area APM will receive a corresponding increment PP′MN, which will be intermediate between the rectangles PR and PN. For the areas of the rectangles PR and PN we shall have

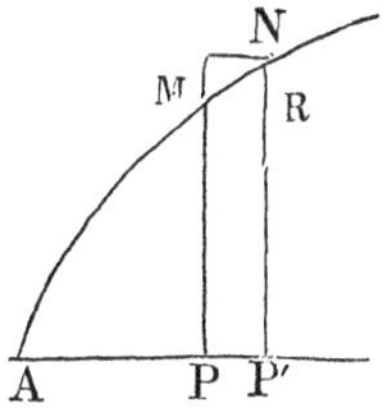

$$PR = yh\ ;\ PN = (y + Ph + P'h^2)h\ ;$$

hence $\dfrac{AP'N - APM}{h} > y$, $\dfrac{AP'N - APM}{h} < y + Ph + P'h^2$,

both of which quantities will, it is evident, be equal to y, when $h = 0$. Thus the limit to the ratio of the increment of the area APM to that of the variable x, on which it depends, will be y, and we shall have

$$\frac{dA}{dx} = y.$$

Let now the equation to the curve be $y = ax^2$. Substituting the value of y we have

$$\frac{d\text{A}}{dx} = ax^2\,;$$

and the question returns to find the inverse function, or integral, corresponding to the differential coefficient $\frac{d\text{A}}{dx}$; or, in other words, to eliminate this coefficient. Applying the necessary rule we obtain

$$\text{A} = \tfrac{1}{3}ax^3 + \text{C},$$

in which the constant will be 0, if we reckon the surface from the vertex A.

2. As a second example, let it be required to find the volume generated by the revolution of a plane curve about a fixed axis.

In the preceding figure let AMN be the generating curve. The volume generated by the small quadrilateral PP′MN will be, it is evident, intermediate between the cylinders generated by the rectangles PR and PN. Employing the same notation as before, we have for the latter,

$$\textit{cylinder}\ \text{PR} = \pi y^2 h\,;\ \textit{cylinder}\ \text{PN} = \pi(y + \text{P}h + \text{P}'h^2)^2 h.$$

Hence, if we put V for the volume sought, we shall have, in the same manner as above,

$$\frac{dV}{dx} = \pi y^2.$$

Let the equation to the curve be $y = ax^2$ as before. Then

$$\frac{dV}{dx} = \pi a^2 x^4\,;$$

whence returning to the inverse function, or integrating, and estimating the volume from the vertex A, we find

$$V = \tfrac{1}{5}\pi a^2 x^5.$$

What has been done will enable the learner to compare the method of Newton with that of Leibnitz. In illustrating the former we have restricted ourselves to the use of the limits or differential coefficients, the proper auxiliaries in this method. We remark in passing that the application of the notation of Leibnitz to the method of Newton, and especially the use of differentials in the latter, serve to confound the two methods, which should always be kept distinct.

DERIVED FUNCTIONS.

194. Departing from the method of Leibnitz and Newton, Lagrange has endeavored to construct the Calculus upon a basis purely algebraic. In order to this he employs as auxiliary to the formation of equations the *Derived Functions* of the quantities under consideration, and which are simple algebraic expressions. The manner in which these functions are derived are, for our present purpose, sufficiently indicated, Alg. art. 237. Thus let $f(x) = ax^3$, for example; the *derivative* will then be $3ax^2$.

195. The derived function $f'(x)$ of Lagrange, the limit L $\frac{\Delta y}{\Delta x}$ of Newton, and the differential coefficient $\frac{dy}{dx}$ of Leibnitz, it will be observed, are precisely one and the same function seen from three different points of view. The three methods are, therefore, fundamentally the same. Each has the same object, that of facilitating the formation of equations by aid of special auxiliary quantities which have certain relations to the primitive quantities for which they are substituted or with which they are used. The office of the Calculus, in either case, is to determine these relations and to provide for the elimination of the auxiliaries when they have served their purpose. As a logical basis of the Calculus the method of Newton, and especially that of Lagrange, may have some advantage. In other respects the superiority is immeasurably on the side of the method of Leibnitz.

....

SECTION XX.

MISCELLANEOUS EXAMPLES.

196. Having thus completed our view of the Calculus, we close with some miscellaneous examples.

Ex. 1. Find the subtangent of the curve whose equation is $xy^2 = a^2(a - x)$.

Ans. $-\frac{2(ax - x^2)}{x}$.

Ex. 2. Find the subnormal of the curve whose equation is $y^2 = 2a^2 \log x$.

Ans. $\frac{a^2}{x}$.

Ex. 3. Investigate the figure and quadrature of the curve represented by the equation $x^4 - a^2x^2 + a^2y^2 = 0$.

Ans. 1°. The curve is included between the limits $x = a$, $x = -a$, and passes through the origin. 2°. The parallels to the axis of y at the distance $x = \pm a$, are tangents to the curve. 3°. There is a multiple point at the origin at which the two tangents are inclined to the axis of x at an angle of 45°. 4°. The whole area is $\frac{4}{3}a^2$.

Ex. 4. Find the area of the curve of tangents, the equation to which is $y = \tan x$. Ans. $A = -r^2 \log \cos x$.

Ex. 5. The equation to the tractrix is $\frac{dy}{dx} + \frac{y}{(a^2 - y^2)^{\frac{1}{2}}} = 0$; required the length of the curve. Ans. $s = a(\log y - \log a)$.

Ex. 6. The Hypocycloid is a curve generated by a point in the circumference of a circle which rolls upon the exterior of the circumference of another circle. Let the equation to one of these curves be $x^{\frac{2}{3}} + y^{\frac{2}{3}} = a^{\frac{2}{3}}$, a being the radius of the fixed circle. Required the length of the curve.

Ans. The curve re-enters after four revolutions, and $s = 6a$.

Ex. 7. Find the volume of the solid generated by the revolution of the curve $y^2 = \frac{b^2x}{a - x}$. Ans. $ab^2\pi \log \frac{1}{a - x} - b^2\pi x$.

Ex. 8. Find the convex surface of the groin, the co-ordinate sections being equal semicircles. Ans. Twice the area of the base.

Ex. 9. If a circle whose radius $= a$ revolve about an axis in its plane, and $c =$ the length of the perpendicular let fall from its centre on the axis; required the volume and surface of the figure generated. Ans. $V = 2\pi^2a^2c$, $S = 4\pi^2ac$.

Ex. 10. Find the arc, less than a quadrant, for which the excess of the sine above the ver-sine is a maximum.

Ans. The arc is 45°.

Ex. 11. Of all right triangles of a given perimeter, to determine that whose area is the greatest.

Ans. That in which the two sides are equal.

Ex. 12. What is the length of the axis of the maximum parabola which can be cut from a given right cone?

Ans. Three-fourths the side of the cone.

Ex. 13. From the extremity of the minor axis of an ellipse draw the longest line to the circumference.

The minor axis of the ellipse being taken as the axis of x, we have for the abscissa of the required line $x=\frac{a^2b}{b^2-a^2}$.

Ex. 14. A line PA is perpendicular to another $AB=a$. Required PA, so that the time of falling down PA + the time of describing AB with the last acquired velocity continued uniform, may be a minimum. Ans. $PA=\frac{1}{2}a$.

Ex. 15. Given the base of an inclined plane $=a$; what must be its altitude, in order that a heavy body may descend down the plane in the shortest time possible?

Ans. It must be equal to the base.

Ex. 16. Find the centre of gravity of a parabolic area, the equation to the curve being $y^2=cx$. Ans. $GG'=\frac{3}{5}x$, $GG''=\frac{3}{8}y$.

Ex. 17. Find the centre of gravity of any parabola whose equation is $y^{m+n}=a^m x^n$.

Ans. $\frac{m+2n}{2m+3n}x$.

Ex. 18. From a solid cylinder of given dimensions cut out a rectangular beam that shall be the strongest possible.

The strength of the beam at any point is measured by the area of its lateral section multiplied by the distance of its centre of gravity from the under surface. Let $x=$ the vertical, $y=$ the horizontal side of the beam.

Ans. $x=\frac{2r}{\sqrt{3}}$, $y=2r\sqrt{\frac{2}{3}}$.

Ex. 19. Find the time in which a paraboloid of revolution whose altitude is h and parameter p, full of fluid, will empty itself through a small orifice, A, in the vertex, its axis being vertical.

$$\text{Ans.} \quad t = \frac{2\pi p h^{\frac{3}{2}}}{3A(2g)^{\frac{1}{2}}}.$$

Ex. 20. The co-ordinate sections of a groin being the semi-cubical parabola whose equation is $y^2 = px^{\frac{3}{2}}$; find the time in which it will empty itself through an orifice in the vertex.

$$\text{Ans.} \quad t = \frac{2p}{A(2g)^{\frac{1}{2}}} h^2.$$

Ex. 21. Find the attraction of a cone upon a particle of matter at its vertex, the force varying inversely as the square of the distance.

Let $h =$ the altitude of the cone, r the radius of the base.

$$\text{Ans.} \quad 2\pi\left(h - \frac{h^2}{(h^2 + r^2)^{\frac{1}{2}}}\right).$$

Ex. 22. Find, for the same law of the force, the attraction of a cylinder upon a point in its axis produced.

Let h, and h' be the distances of the attracted point from the upper and lower bases of the cylinder respectively, $a =$ the radius of the base of the cylinder; and put $(a^2 + h^2)^{\frac{1}{2}} = c$, $(a^2 + h'^2)^{\frac{1}{2}} = c'$.

$$\text{Ans.} \quad 2\pi(h' - h - (c' - c)).$$

www.ingramcontent.com/pod-product-compliance
Lightning Source LLC
LaVergne TN
LVHW050517100826
845148LV00002B/370

9781425520793